A Practical Guide to Adopting BIM
in Construction Projects

A Practical Guide to Adopting BIM in Construction Projects

Bimal Kumar

Whittles Publishing

Published by
Whittles Publishing,
Dunbeath,
Caithness KW6 6EG,
Scotland, UK

www.whittlespublishing.com

978-184995-146-3

Printed by Gomer Press Ltd.

Contents

Foreword

Since the UK Government Construction Strategy was launched in 2011 with its ambitions for BIM and the beginnings of a digitised construction sector there has been significant progress made in developing a wrapper to enable this collaborative approach. Clients and supply chain organisations are starting to use their data to better design, procure assets and create a more efficient built environment that is more sustainable with better places to live and more intelligent infrastructure.

As industry moves from BIM mobilisation to implementation it is essential that there is practical information at hand to help them successfully execute digital projects and understand how to instil Level 2 BIM across their business. This book offers real-world advice on how you can create that digital transformation and successfully realise the benefits that BIM will bring through more efficient and innovative ways of working.

It's not a matter of if, but when your firm will implement BIM. Are they on the right track and are they moving fast enough?

David Philp FRICS, FCIOB
Head of BIM, UK BIM Task Group

Preface

Building Information Modelling (BIM) is currently the most talked about term within the construction industry all over the world. The UK government has made the use of BIM technologies and processes mandatory on all public sector projects from 2016 regardless of size. Inevitably, this has given rise to a high level of interest in BIM within the UK and elsewhere.

Whilst BIM technologies have been used in a 'lonely' mode for a few years now in several countries, there is generally a lack of proper understanding and appreciation of processes, standards and protocols that need to be in place before a more holistic implementation can take place potentially benefitting all stakeholders in the procurement and operation & maintenance of an asset. There have been innumerable books, papers and reports published on the theoretical underpinnings of BIM technologies but relatively few have dealt with the more mundane practical issues around implementing BIM technologies and processes in a construction project.

This books aims to fulfil that need. Most of the material is country or technology-agnostic although some material may seem to be UK-centric. Even those parts of the book which may appear to be UK-centric will be relevant for any other country and could well be easily adapted. This book attempts to be of practical use to practising engineers, architects, contractors as well as client organisations. However, it will also be of use to students of more advanced built environment courses. The book is organised into seven chapters and four appendices. They are organised purposely in such a way that any one of them could be read almost entirely on its own. However, it is recommended that the first four introductory chapters (1–4) should be read before embarking upon the other chapters individually. The first four chapters should hopefully help the reader develop an understanding of the main drivers and background behind BIM-driven asset life cycle management. In the process, it is hoped that a number of misconceptions and myths about BIM will be addressed helping the reader develop a proper understanding of the key issues and put the whole BIM story into perspective. The four appendices provide some very practical advice and material for implementation of BIM in projects.

The central message of this book is that BIM is all about seamless information *management* for the entire life cycle of an asset rather than simply information *modelling* at the design and construction stages.

Bimal Kumar

1

Introduction and background

1.1 Introduction

BIM (Building Information Modelling) is an acronym that many people in the construction industry are becoming increasingly interested in. BIM is not a new concept, but the current frenzy within the United Kingdom construction industry is driven by the UK Government's directive on using BIM in all public sector projects from 2016.

This book is intended to act as a practical guide for a typical practitioner in the industry. Of late, there has been an explosion of publications, CPD (Continuing Professional Development) courses, seminars and other forms of dissemination activities on BIM in the UK. Therefore, this book does not intend to replicate what has been said or written umpteen times before on the subject. Instead, it aims to be a 'useful' document that can provide guidance on how to deliver BIM-enabled projects. Much of what is dealt with here should be considered as a starting point before embarking on a typical BIM-based project. However, to put things into perspective, a certain amount of basic introductory material and background to the BIM 'story' will be provided in this chapter.

1.2 Common myths about BIM

Right up front, here are some commonly held misconceptions:

- BIM is a piece of software
- BIM will save my firm 20%+ if I use BIM software
- A client is asking me to use BIM on their project – let's buy some Revit (BIM software) licences
- We have been doing BIM for at least 30 years ever since CAD (Computer-aided Design) came out
- BIM is CAD by another name

Suffice to say that at this stage, these are all incorrect in more ways than one. A more comprehensive analysis of these 'myths' will be presented in the next chapter when a detailed discussion on what is and, more importantly, what is NOT BIM, is provided. However, it is first important to consider the nature and characteristics of the construction industry itself, with a view to introducing BIM into asset life cycles.

1.3 Key characteristics of the UK construction industry

The construction industry in the UK (and also most countries in the world) is a major contributor to the economy, accounting for almost 10% of the country's gross national output (ONS, 2010). However, during the recent financial crisis, this dropped to approximately 7%. The industry employs some 1.5 million people directly, and considerably more indirectly, accounting for about 4.5% of the employed labour force in the UK (ONS, 2010). Despite its relative importance for the UK economy, this is an industry fraught with several deep-rooted problems that go back to its origins in being a craft-based industry. This is a project- rather than a product-based industry, which means that it has a very unique set of characteristics that are often quite complex. For example, the key stakeholders in any construction project are 'forced' to form short-term relationships and collaborate in order to deliver a project successfully. This requires a very responsive and agile approach to working with relatively new partners and developing working relationships quite quickly. By the time these relationships have cemented, the project may well be close to completion, when the stakeholders move on to another project, only to repeat the process with perhaps a totally new set of partners! This is not easily accomplished and yet the success of the industry depends on it. Therefore, the industry needs to have appropriate infrastructure in place for effective partnerships. One way to achieve this is to have effective information exchange strategies in place. However, in an

How the customer explained it

How the project leader understood it

How the analyst designed it

How the programmer wrote it

How the business consultant described it

How the project was documented

What operations installed

How the customer was billed

How it was supported

What the customer really needed

Figure 1.1: Poor information management (reproduced from www.projectcartoon.com)

industry predicated upon people from different organisations collaborating for short periods of time, it is fraught with extremely complex information flows. In order to address this effectively, one of the key ingredients of the required infrastructure to facilitate smooth information flow is the use of standards and protocols for effective information creation, storage, exchange and management. As this is not in place on most projects, the result is that the project outcomes often are not satisfactory, and clients are frequently left accepting an asset that they did not ask for. The other hugely important negative impact of all this is that the industry as a whole ends up with adversarial relationships between the key stakeholders, resulting in claims and counter claims. The industry, rather perversely, exploits the incomplete and fuzzy nature of information that it has to work with by making most (if not all) of its profits from claims and counter claims. Figure 1.1 illustrates this issue of poor information management leading to an outcome that was totally unintended and different from that envisioned to start with.

This section has presented a very brief overview of the nature and characteristics of the construction industry because a more detailed one is outside the scope of this book. For interested readers, there are several other excellent publications (e.g. Crotty, 2012) that discuss these issues in much greater detail.

1.3.1 Role of information exchange in the construction industry

In order to appreciate the importance of information exchange within a typical construction project, figures 1.2 and 1.3 show a few examples of the enormous volumes of information generated during the course of various kinds of typical construction projects. Figure 1.2 is a snapshot of results from analysis of EDMS (Electronic Document Management System) for a PFI (Private Finance Initiative) hospital project. Figure 1.3 shows similar results for a JV (Joint Venture) motorway extension project.

It is plainly obvious that without effective and sophisticated standards, protocols, processes and technologies in place, it will not be possible to exchange such volumes of information effectively in a seamless manner. If the information is not managed effectively, the side effects are all too familiar to anyone involved with the industry: significant time and money lost on claims and other forms of litigation, as well as substantially escalated costs and schedules of projects.

1.4 Issues of interoperability within the construction industry

As discussed in section 1.3, the role of high-quality information is vital for successful delivery of a construction project. This includes efficient methods of generation, storage, sharing and management of information. One of the key requirements for this is effective and good interoperability of information between systems used by different stakeholders in a project. However, it is a matter of common experience and knowledge that interoperability is a major issue in the construction industry, not just in the UK but also worldwide. Most systems used in the industry do not ‘talk’ to each other. A major study carried out by NIST (National institute of Standards and Technology) in

Communication Type	# created	# attached	# per month
Builder's work request	90	100	3
Change control proposal	1500	500	50
Client observation record	200	250	7
Company record	450	0	15
Contact record	1300	0	43
D&B contractor's instruction	1500	350	50
Decisions made	1500	3	50
Directive	1500	0	50
Drawing record	35000	0	1166
Fax	3300	1650	110
Impact analysis form	4200	900	140
Incoming documents	8300	8100	277
Letter	2000	600	67
Electronic memo	8300	6400	277
Minutes	2000	1500	67
Site direction	150	10	5
Site instruction	10	10	0.3
Sign-off notification	550	1115	18
Telephone call notes	100	10	3
Technical query	4000	2550	134
Transmittal	16000	200	534
Totals	**91950**	**24248**	**306**

Figure 1.2: Volume of information generated in typical projects (from Sommerville and Craig, 2002)

Communication Type	# created	Communication Type	# created
Telephone notes	78	Site instructions	7
Meeting minutes	1855	Final design drawings	2094
Electronic memos	2691	Method statements	122
Letters	7157	Non-conformance reports	3191
Incoming documentation	18329	Project certificates	7000
Faxes	3076	Quality procedures	111
Approval to proceed	20	Requests for inspection	2262
Confirm verbal instruction	3	Site photos	408
Engineer's instructions	1	Technical queries	1670
Instructions to subcontractors	1276	Project contacts	3000
Requests for information	11		
Subtotals	**34497**		**19865**
Totals	**54362**		

Figure 1.3: Volume of information generated in typical projects (from Sommerville and Craig, 2002)

the United States of America (Gallaher *et al.*, 2004) concluded that the cost of lack of interoperability between systems used in this industry was well over US$15 billion per year! A further result of this study was that 68% of this was incurred (US$10.8 billion) in the O&M (Operation and Maintenance) stage. Although these are USA figures, it is likely that the figures for the UK and other comparable developed economies will be proportionately similar. Therefore, the lesson is that the industry has to strive to reach a stage where systems can talk to each other as seamlessly as possible, thus saving the industry very substantial sums of money. This issue of interoperability will be considered again in chapter 4 when different possible solutions to this problem will be discussed.

1.5 Common problems in the construction industry

Fragmentation in the industry:

- Majority of the companies in the construction industry in most industrialised nations consists of less than five people. It is hard to find the kind of investment required for productivity increases through effective collaboration.
- Short-term relationships between 'collaborating' firms in a typical project means there is a lack of sufficiently long engagement to form standard practices for communications.

According to Crotty (2012, p. 26) 'Unpredictability and low profitability are both caused by the same underlying phenomenon in large part – the devastatingly low quality of most of the information used on modern construction projects'.

Another key characteristic of this industry that leads to unsatisfactory outcomes is the lack of a 'holistic' view by the entire project team of the life cycle of the asset they may be building. What this means is that typically the design team normally works in isolation from the rest of the stakeholders, most notably those who are responsible for the asset after construction and handover, i.e. asset and facility managers. This often leads to several issues that the designers do not even consider, whereas some of these issues could be very simple and straightforward to address if the facility managers were involved in the design stage. There are several constructability issues that would be relatively insignificant to tackle at an earlier phase if contractors were involved in the process right at the start of the life cycle. The same simple issues become hugely complicated (and consequently expensive) to fix if left until later. The MacLeamy curve in figure 1.4 and the diagram produced by the BIM Task Group (figure 1.5) illustrate this point very clearly.

Figure 1.6 is an attempt to capture the track record of the construction industry on megaprojects. The pie chart on the right gives a breakdown of the possible causes of failures on large construction projects. It is clear from the pie chart on the left that the vast majority of large projects tend to underperform, whereas the chart on

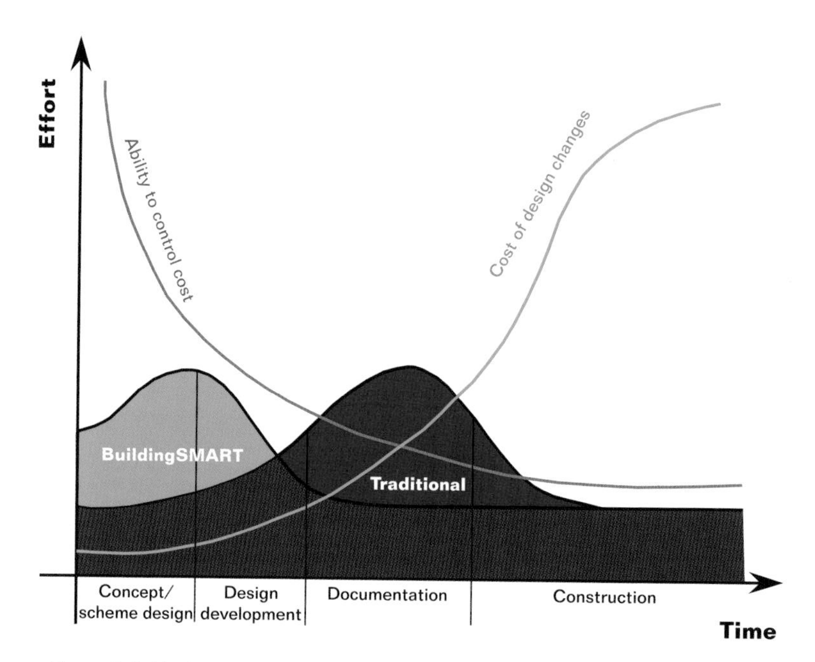

Figure 1.4: MacLeamy curve

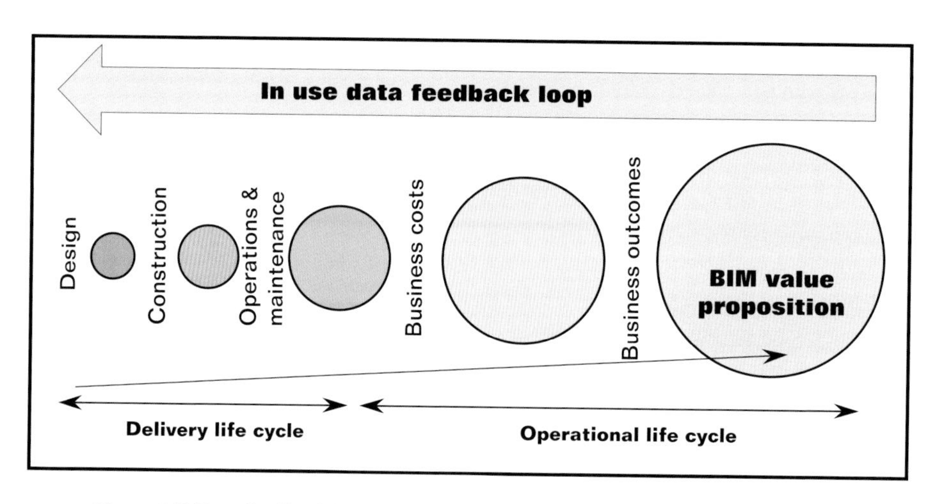

Figure 1.5: Data feedback loop (redrawn with permission from BIM Task Group)

the right suggests that only a small proportion of the failures are due to technical difficulties and that the vast majority of failures are due to non-technical issues. The three largest categories in this chart are Poor Organisation and Project Management Practices, Poorly Defined or Missing Project Objectives and Ineffective Project Planning. Between them, they constitute some 71% of the causes of project failures.

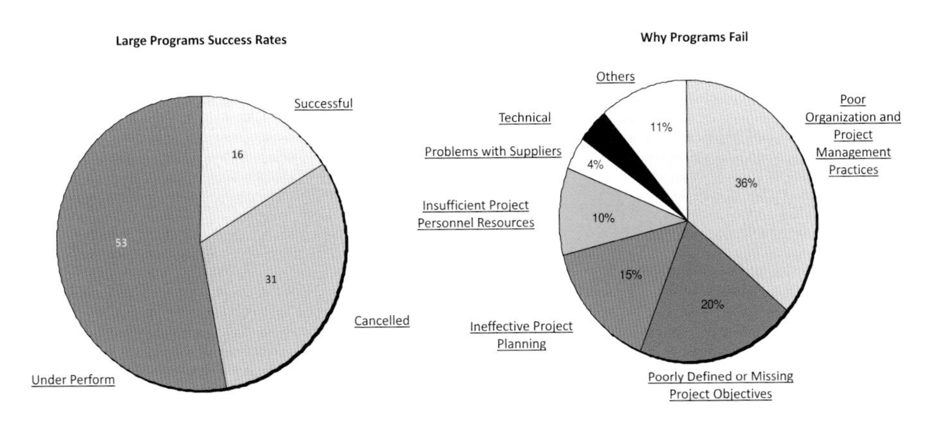

Figure 1.6: Failure on mega projects (Kennerson, 2013)

Interestingly, one can easily attribute all these three categories to lack of high-quality information management in one way or another.

1.6 Latham and Egan reports

In view of the above-mentioned issues with the construction industry, the UK Government has set up various committees over the years to investigate the causes behind these issues and to propose solutions. Among the key reports produced in the recent past, two of the most important are: *Constructing the Team* (Latham, 1994) and *Rethinking Construction* (Egan, 1998).

Latham's report identified several industry inefficiencies, citing the fragmented nature and adversarial model of project delivery prevailing in the industry as two of the main reasons behind them. He proposed 53 recommendations mostly centred around effective teamwork and collaboration between key project stakeholders. It was this report that put forward ideas such as partnering and introducing greater use of information and communication technologies.

Egan's report was published four years after Latham's and was an equally, if not more, influential report in the UK construction industry. The Egan report identified five key drivers of change in order to implement an integrated project:

- committed leadership
- focus on the customer
- integrated processes and teams
- quality-driven agenda
- commitment to people

Many years after the publication of these two influential reports, the industry is still grappling with the same issues. There are several reasons for this, but it would probably not be too critical to suggest that both reports targeted only parts of the story. They focused on people issues and tried to suggest methods to reduce adversarialism and confrontation between key stakeholders. Whilst this was relevant, they missed out on some key aspects of how to address these challenges, and arguably neither report actually made any significant impact on the industry's performance.

The approach of the two reports for solving the issues of confrontation was to focus on contractual and legal aspects of project delivery. It is suggested here that whilst this may have been a perfectly reasonable approach, it missed out on the key driver and source of most (if not all) of these confrontations. It is reasonable to suggest that many of these confrontations arise because of lack of high-quality information and effective means of communicating it between all stakeholders. BIM aspires to address exactly this challenge.

Both Latham and Egan picked up on relatively low usage of IT (Information Technology) tools within the industry to communicate information. However, both fell short of addressing the challenge of how information was generated, stored and communicated. It is common knowledge that technologies are mere facilitators, but unless all stakeholders speak the same language, meaning that they all use the same standards and protocols and processes for information generation and communication, even the best state-of-the-art technologies will not be effective. This is no different from expecting a confrontational group of people who all come from different linguistic backgrounds speaking different languages to resolve differences by communicating in their own language using the best technologies available! Clearly, both the Latham and Egan reports, which were valuable for several important insights into the industry's functioning, failed to change the industry for the better, with the result that the industry still fundamentally works no differently from several decades ago. In fact in this sense, the two reports did not conclude anything significantly different from dozens of earlier reports, as Murray and Langford (2003) point out in their book, which presents analyses of several such reports published since 1944.

1.7 Low profit margins and profitability in the construction industry

As mentioned earlier, the construction industry makes substantial sums of money by exploiting the fuzziness and lack of information through claims and counter claims. This is also driven partly by the fact that this industry suffers from very low profit margins, which is, in turn, driven by the fact that contractors bid low to win a contract and often rely on claims later on to make any money. This is clearly not a healthy situation for any industry and needs to be addressed. However, to be fair, the reason many contractors end up in this kind of scenario is that the available information base on which they bid for jobs is not of sufficient high quality. In this section, a brief analysis is presented on adjudication-based claim settlements in order to put things into perspective. Kennedy *et al.* (2010) present a detailed analysis of adjudication cases, which is one of the two main mechanisms used in the industry to settle claims; the other mechanism is arbitration but any data on it is much more difficult to obtain as it is held in tightly controlled secrecy. Table 1.1 (Kennedy *et al.*, 2010) gives a breakdown of data on the parties most frequently falling into dispute between October 2005 and April 2008 in the UK. It clearly shows that the most common pair falling into dispute is the main contractor and the subcontractor, followed by the client and the main contractor. Interestingly, the nominated subcontractors have disappeared altogether and trade or package contractors have increased almost certainly as a result of changing procurement routes used by clients over the period in question.

Table 1.1: Data on the parties most frequently falling into dispute (from Kennedy *et al.*, 2010)

Contracting Parties	October 2005 (%)	October 2007 (%)	April 2008 (%)
Client and consultant	3	5	4
Client and main contractor	35	27	37
Client and nominated subcontractor	1	0	0
Main contractor and domestic subcontractor	51	52	47
Main contractor and nominated subcontractor	2	0	0
Sub-contractor and sub sub-contractor	3	8	4
Consultant and contractor	1	2	1
Trade contractor and employer	2	2	4
Management contractor and package contractor	2	3	4
Consultant and sub-consultant	0	0	1

Table 1.2 shows the adjudicators' views on the main causes of disputes between July 2004 and April 2008. These factors are clearly subjective and may not be a true reflection of all the issues involved.

Table 1.2: Data on the main causes of disputes (from Kennedy *et al.*, 2010)

Subject	July 2004 (%)	October 2005 (%)	October 2007 (%)	April 2008 (%)
Valuation of Final Account	12	14	22	22
Failure to comply with payment provisions	19	14	8	19
Valuation of interim payments	15	13	15	16
Withholding monies	10	11	10	10
Extension of time	8	8	8	9
Loss and expense	9	10	2	7
Valuation of variations	15	17	11	5
Defective work	4	5	7	4
Determination	2	3	4	4
Non-payment of fees	2	1	7	2

Table 1.3: Trend in adjudications in the UK (from Kennedy *et al.*, 2010)

Time Periods	All ANBs Adjudications
YEAR 1 - May 1998 - April 1999	187
YEAR 2 - May 1999 - April 2000	1309
YEAR 3 - May 2000 - April 2001	1999
YEAR 4 - May 2001 - April 2002	2027
YEAR 5 - May 2002 - April 2003	2008
YEAR 6 - May 2003 - April 2004	1861
YEAR 7 - May 2004 - April 2005	1685
YEAR 8 - May 2005 - April 2006	1439
YEAR 9 - May 2006 - April 2007	1506
YEAR 10 - May 2007 - April 2008	1432
YEAR 11 - May 2008 - April 2009	1737
YEAR 12 - May 2009 - April 2010	1528

Finally, table 1.3 (Kennedy *et al.*, 2010) shows the trend in adjudications in the UK over the period May 1998 to April 2010. The table clearly demonstrates a rapid increase in the number of adjudications in the first two years after adjudication was introduced into the UK, then it remains constant for two years, before falling and settling at a new level. It should be pointed out that this is not the full extent of disputes because there are other routes to dispute resolution that people use, and it is difficult to obtain a full picture without data on all the options taken. This is a sensitive matter, and a full set of data on all disputes is not likely to ever be available in the public domain.

Considering the volume of claims and other associated issues, it is not entirely unreasonable to conclude that the industry has a vested interest in maintaining the status quo! This is not surprising as most organisations resist any kind of change, particularly the ones that at least ostensibly appear to cut off a good source of profitability!

1.8 Organisation of this book

The rest of this book is organised into six further chapters. Here is a brief overview of these chapters. The chapters are organised purposely in such a way that any one of them could be read almost entirely on its own. However, it is recommended that everyone should read the first three introductory chapters (2–4) before embarking upon reading the other chapters individually.

Chapter 2 deals with the basics of information modelling and management principles. Although parts of this chapter are theoretical, the reason it has been included here is to provide a greater appreciation of the BIM technologies covered in later chapters. This chapter could well be skipped, but it is not recommended. It also covers the importance of information management in construction in some detail.

Chapter 3 provides an overview of the UK Government's BIM initiatives. This is the first chapter where more UK-specific issues start being covered. The preceding chapters are country-agnostic. However, it should be pointed out that the UK

Government's BIM strategy is quickly gaining acceptance and adoption in other parts of the world, and in this sense, the whole book can treated as country-agnostic. This chapter covers in some detail the key concepts driving the UK Government's initiatives.

Chapter 4 provides a whirlwind tour of the most important and biggest documents of all, PAS 1192:Part 2 and EIRs (Employer's Information Requirements). The idea is to provide an analytical view of the documents and to identify the key parts that a practitioner needs to have a clear understanding of. There are several pieces of guidance and helpful advice provided for the practising person, thus making the process of using this publication in a project so much easier.

Chapter 5 provides guidance on producing a BIM PEP (Project Execution Plan) document. As BIM PEP documents are a key element of any BIM-enabled project, a methodology for developing these documents is provided in some detail.

Chapter 6 deals with three other publications: the BIM protocol, the Outline Scope of Services and PII (Professional Indemnity Insurance). As these documents are relatively short, they are all combined into one chapter.

Chapter 7 provides an overview of training-related issues for the federated Level 2 BIM and suggests some guidelines for developing a BIM training strategy for different kinds of organisations.

The four appendices provide valuable practical pieces of information of direct relevance to BIM implementation in projects. For example, in appendix 3 there is a sample BIM PEP document provided from an organisation that is using it in live projects. The other appendices include a guideline for level 2 compliance, typical responses to a standard preliminary qualification questionnaire (PQQ) in relation to BIM and a sample training course curriculum based on BIM Task Group's Learning Outcomes Framework (LoF).

1.9 Summary

In summary, the construction industry suffers from several issues that can cause waste and lack of profitability. Some of the key issues behind this are adversarial relationships and lack of effective collaboration underpinned, at least in large part, by a lack of high-quality information capture and management processes and technologies. Therefore, the construction industry's record is relatively poor and characterised by inefficiencies, adversarial relationships and low profit margins. Actually, it also has a poor sustainability record and is the only sector out of ten major sectors in the UK that increased its greenhouse gas emissions per unit of output between 1990 and 2008! A mixture of process, technology and people issues appear to beset this industry, inhibiting it from becoming a thriving, innovative and profitable one. If the industry is to make that leap, it needs to sort out all these issues, which is where BIM processes and technologies become important, as discussed in the following chapters.

2

Information management for the construction industry

2.1 Introduction

As discussed in chapter 1, it is clear that information management plays an important role in the successful delivery of projects in the construction industry, and that a typical construction project involves enormously complex information flows. It is probably not an exaggeration to say that most major problems of this industry stem from the highly complex and unstructured nature of information flows and exchanges in a typical construction project. This is a problem that does not stand isolated from a number of other related problems which must be addressed before one can hope to achieve the kind of overall efficiencies that several major studies have alluded to. To summarise succinctly, it would be apt to paraphrase a certain US president, 'It's all about information, stupid!!'.

This chapter presents a brief overview of information modelling and management concepts, and related issues, before presenting an introduction to BIM and its relevance to project delivery.

2.2 Challenges with information flow in the construction industry

As discussed earlier, the fragmented nature of this industry generates so many issues that several other comparable industries just do not have to grapple with. In terms of key challenges regarding information flow and exchange in the construction industry, the following are some of the main issues:

- The vast majority of the industry comprises micro firms, who cannot afford the infrastructure or expertise for seamless information exchange with other stakeholders in a project.
- No single organisation has the financial muscle power and influence to enforce industry-wide standards for information exchange. In most markets, no single organisation in this industry has more than a low single-digit market share, unlike several other industries such as high-technology industries (for example) where Intel had an 80%+ share of the microprocessor market (Digikey, 2011) in 2010.

- Linked with the point above, there is a lack of agreed standards for information exchange that cover all aspects of the architecture, engineering and construction industry.
- The discrete nature of project delivery processes with disjointed stakeholders gives rise to a lack of seamless co-ordination and collaboration.
- The constantly dynamic relationships between stakeholders due to the project- rather than product-based nature of the industry creates complexities.

In light of these factors, it is proposed that rather than project management simply being 'the planning, organising, monitoring and control of all aspects of a project and the motivation of all involved to achieve the project objectives safely and within agreed time, cost and performance criteria', the industry should consider it fundamentally as an *information processing enterprise* with its ultimate goal being that:

> All stakeholders in any project share and exchange information in a pre-agreed standard format that is enabled by appropriate technologies, processes and standard protocols.

2.3 Model-based Design

As a prelude to developing the case for BIM and to put things into perspective, this section first provides a brief discussion of MBD (Model-based Design), with a view to bringing out the benefits of this approach over the more traditional drawing-based design. MBD is a mathematical and visual method of addressing problems associated with designing complex systems such as buildings, plants or indeed any other product. Among other things, MBD provides an efficient and cost-effective approach for facilitating seamless communication and hence integration throughout the design and construction processes.

Recently, developments in Internet technologies have facilitated MBD approaches by allowing retrieval of design-related information available on the Internet and seamlessly integrating distributed applications as web-based services within CAD systems. Figure 2.1 (Cheng *et al.*, 2010) shows a conceptual framework for web-enabled model-based CAD applications. This approach allows for incorporating the most up to date information in a design, whether that relates to individual elements of the design or any other design aspects such as perhaps the latest regulatory information. Cheng *et al.* (2010) present this approach to MBD, which links a design system with an online *catalogue* of products that can be simply *dropped* into the design model with some simple mouse clicks. Once the products or elements are dropped into the models, they can then be manipulated as if they were an integral part of the original model. Several design software vendors have produced such technologies (e.g. Autodesk's i-Drop or

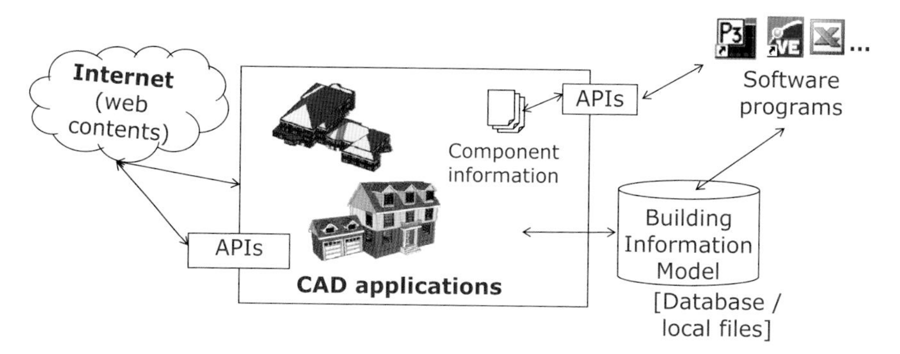

Figure 2.1: Web-enabled model-based CAD applications (reproduced from Cheng *et al.*, 2010)

Google's 3D Warehouse). This makes the communication and exchange of information between different designers and manufacturers so much more seamless, timely and effective. Further downstream it is then possible to carry out parametric and other analyses by manipulating these products by interchanging them with perhaps other manufacturers and choose the best possible solution for the client.

Cheng *et al.* (2010) present an example of such analysis for the energy performance of a building by replacing the door and windows of a building and in a relatively short space of time comparing their energy and carbon footprints. Figure 2.2 shows the model-based design system interacting with the web-based *catalogue* (3D Warehouse in this case) before choosing and dropping a particular type of window into the design model. Figure 2.3 shows that the energy analysis system first extracts relevant information from the design model before carrying out the analysis. Finally, figure

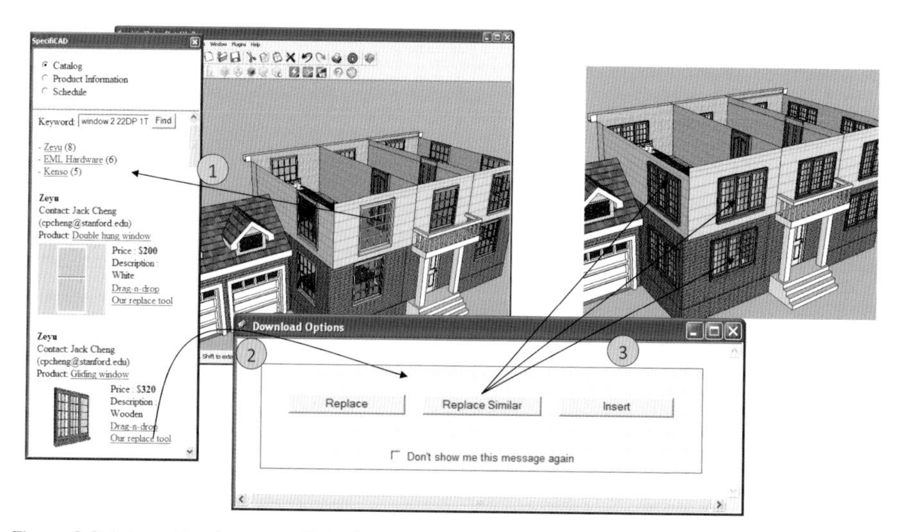

Figure 2.2: Interaction between CAD object components and dynamic online information (reproduced from Cheng *et al.*, 2010)

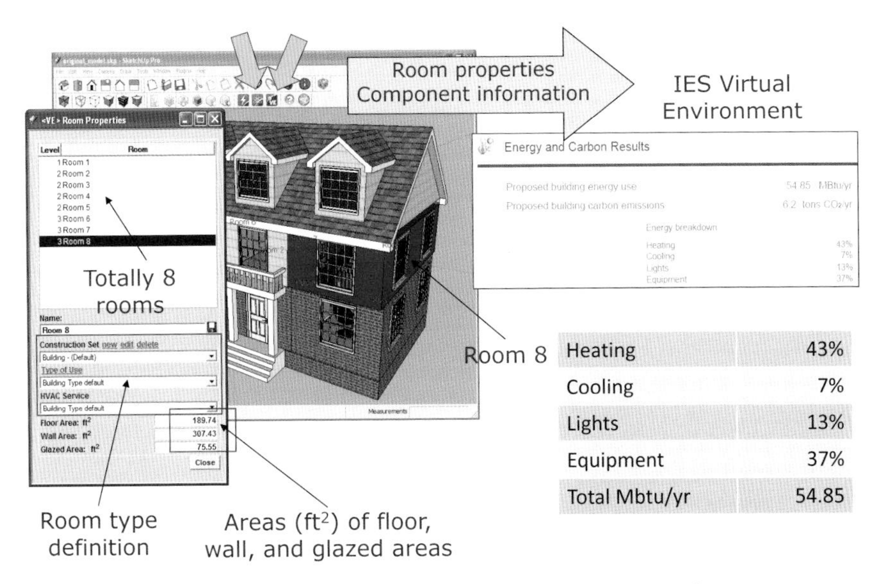

Figure 2.3: Energy and carbon emissions analysis of a building (reproduced from Cheng *et al.*, 2010)

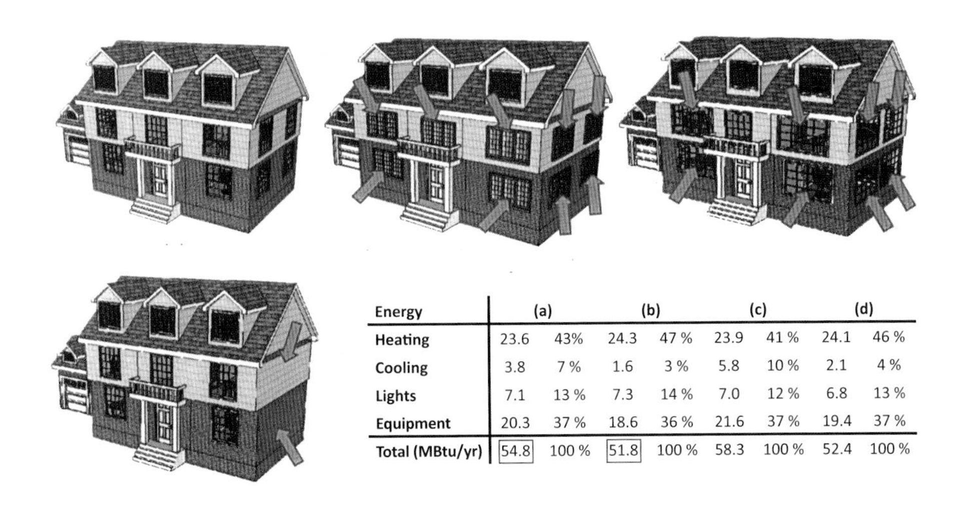

Energy	(a)		(b)		(c)		(d)	
Heating	23.6	43%	24.3	47 %	23.9	41 %	24.1	46 %
Cooling	3.8	7 %	1.6	3 %	5.8	10 %	2.1	4 %
Lights	7.1	13 %	7.3	14 %	7.0	12 %	6.8	13 %
Equipment	20.3	37 %	18.6	36 %	21.6	37 %	19.4	37 %
Total (MBtu/yr)	54.8	100 %	51.8	100 %	58.3	100 %	52.4	100 %

Figure 2.4: Comparison of different architectural designs in terms of energy use (reproduced from Cheng *et al.*, 2010)

2.4 shows the impact of changing the window types on the energy consumption and carbon emissions of the building.

It should be stressed that using the traditional CAD-based design systems, this kind of energy analysis would not be possible without a significant amount of programming and other technical customisations. This would not just make the whole process cumbersome but extremely tedious, requiring a significantly higher amount of time that would render the whole effort infructuous.

2.4 BIM

So, with the background provided in chapter 1 and a very brief introduction to MBD earlier in this chapter, it is now an opportune time to introduce BIM. However, it should first be mentioned that one of the problems with the whole BIM 'thing' is that BIM technology hit the scene way before BIM processes, protocols and standards. The upshot of this is that someone familiar with BIM technologies takes it as such, and no more. In fact, it would arguably be more appropriate if the 'B' in BIM should stand for 'Building' as a verb rather than a noun, implying its scope to be vertical (building) structures as well as horizontal (civil infrastructure) structures such as bridges, tunnels, etc. Similarly, the 'IM' in BIM should more appropriately stand for 'Information Management' rather than 'Information Modelling'. The wider connotation of BIM is much bigger than just modelling and has huge implications for overall information management for the entire asset life cycle from inception to demolition. This includes modelling, but is not restricted to it. In fact, many of the misconceptions about BIM arise more because of this historical fact than anything else. Therefore, BIM normally means not just the

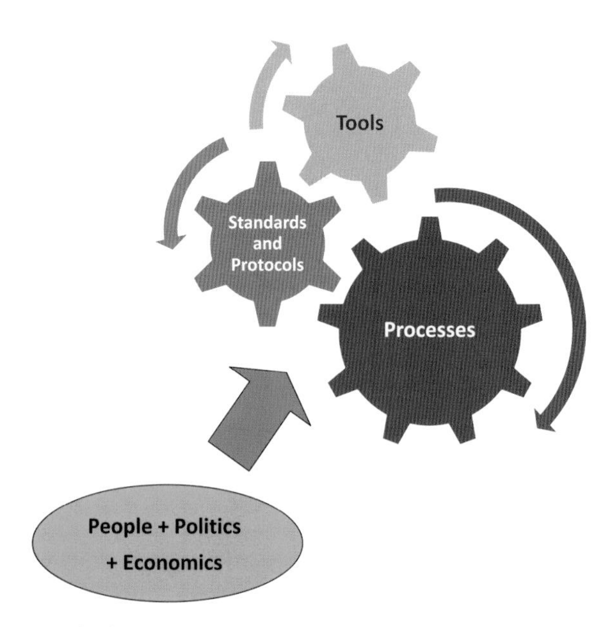

Figure 2.5: Interplay between BIM processes, standards, technologies and people

technology but also the associated set of processes, standards and protocols for information exchange. All these aspects of BIM have to work together, as illustrated in figure 2.5. However, as the diagram also suggests, despite tools (technologies), standards and processes working perfectly, the biggest influence, as always, is that of people, economics and political issues, which may make or break a perfectly good and effective way of procuring projects using BIM.

The rest of this chapter first introduces the basic concepts of information modelling itself before going into details of BIM technology. Later chapters will deal with processes, standards and protocols, which must be embedded in the procurement process facilitated by the BIM technologies presented in this chapter.

2.5 Data, information and knowledge

As the core concept this book deals with is *information,* let us first consider the *data-information-knowledge* continuum. So, how does one define information and what is the difference between data, information and knowledge? For centuries, philosophers have grappled with these terms. However, the discussion here will be limited to its relevance to the core topic being covered: BIM. The simplest way to define *data* is that *it is a collection of symbols randomly put together that does not convey any meaning. Information*, on the other hand, is *data organised in a way that conveys a meaning.* Finally, *knowledge* could be defined as *information with the addition of a person's own world view giving the meaning a unique insight for that person.* For example, figure 2.6 on the left shows a collection of zeroes and ones that to most people conveys nothing meaningful. Therefore, this is simply data for most people. However, to a trained eye (a computer scientist or a mathematician), it conveys the word 'Wikipedia' in binary code. This is clearly *information* rather than just *data* to those people who understand such code. Finally, the word 'Wikipedia' will convey several different *meanings* to different people depending on their own personal world view. This is where one may have transcended from *information* to *knowledge.*

There are several books written on epistemology covering these very complex issues, but all are well outside the scope of this book. For interested readers, Bronowski (1978) is a good source of material on these issues.

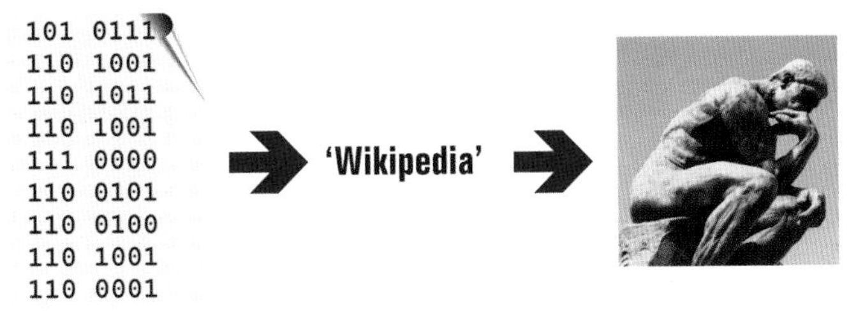

Figure 2.6: Data ➔ information ➔ knowledge continuum

2.6 Information modelling

An information model in software engineering is a representation of concepts and the relationships, constraints, rules and operations to specify data semantics for a chosen domain of discourse. Typically, it specifies relationships between kinds of things or objects, but may also include relationships between individual things.

The term *information model* in general is used for models of individual items, such as facilities, buildings, process plants, etc. In those cases, the concept is specialised to the facility information model, building information model, plant information model, etc. Such an information model is an integrated *view* of the facility that the data and documents about the facility relate to.

2.7 Entity–Relationship-based information modelling

One of the most common and popular approaches to information modelling is known as E–R (Entity–Relationship) modelling (Elmasri and Navathe, 1994). Again, a very brief and simple overview is provided here simply to put things into perspective. Entities are akin to objects (physical or conceptual), and this modelling approach relies on identifying relationships between all the entities inhabiting the universe of discourse (e.g. a building) under consideration. Additionally, each entity or object is defined by a set of properties.

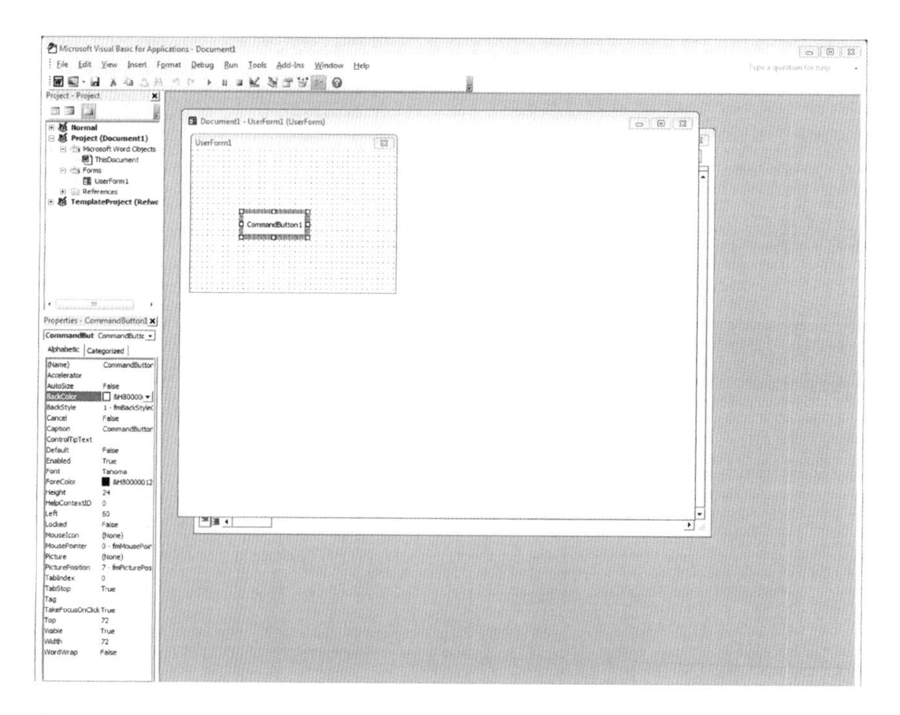

Figure 2.7: Development environment GUI for Microsoft Word (source: Microsoft Corp.)

Figure 2.8: Popular BIM system (Revit) GUI (source: Autodesk Inc.)

As will be shown later with a simple example, it is then possible to reason about the objects in quite clever ways that may not have been possible without modelling them in this way. BIM technologies utilise exactly the same approach.

Many other popular commercial systems also utilise this approach. This can be demonstrated by examining the GUIs (Graphical User Interfaces) of these systems and the striking degree of similarity between them despite being developed for radically different purposes! For example, the GUIs of a popular word processing software and a BIM software are shown in figures 2.7 and 2.8, respectively.

The similarity is striking. The two smaller windows on the left have even got the same name (*Properties* and *Project*). The window on the right is where one does all the *work* (i.e. modelling in a BIM system or programming some customised functionality in the other software). The *Properties* window displays all the properties of the selected object in the window on the right (i.e. properties of the wall (say) such as thickness, area, etc., in the BIM software and properties of the *messagebox* (say) such as its caption in the other software). Arguably, it is useful to appreciate this very basic technical background for the *genre* of software that BIM belongs to simply to understand the background of the technology being used.

So, in summary it can be said that:

- BIM technology is based on object-based information modelling

- the model is organised in terms of the hierarchy of objects (a bit like building–rooms–people–etc.)
- each object contains parameters (which could be pre-defined or user-defined at the time of use) and a set of rules about its behaviour
- this kind of organisation of information about the objects makes it relatively straightforward to retrieve information about their different parameters as well as reasons about their behaviour

In addition, BIM technologies can:

- use pre-defined base (standard) objects
- allow users to define new objects
- allow users to alter properties of existing objects
- have the ability to store all relevant information about a building in one place
- facilitate extraction of useful information about different aspects of the building (and its components) from a single system

On the other hand, traditional CAD technologies:

- essentially put together a collection of primitive geometric shapes using a couple of different approaches (outside the scope of this book)
- some of the more recent 'modern' CAD systems do use object-based approaches; however, they differ from BIM technologies in several ways, the most important of which are:
    - the lack of facility for user-level object creation and object parameter value updates
    - parametric modelling tools move up the value chain by transforming the geometric design tools to information or knowledge-based design tools

Now, a brief look into some of the key characteristics of object-based parametric modelling is provided. Properties or parameters that identify with a *class* of objects are defined only once – with the class – and are *inherited* by all objects (instances) of the class, but the objects can have additional, specific properties and methods (rules) that are not shared by its parent class (e.g. see figure 2.9).

A *class* is akin to a *family* in a typical BIM system, e.g. the wall or column family. Therefore, whenever a wall is created in a BIM system, the basic parameters or properties of the wall family are inherited by it and the values for these properties are assigned by the user. The advantage of this approach is that all the properties do not

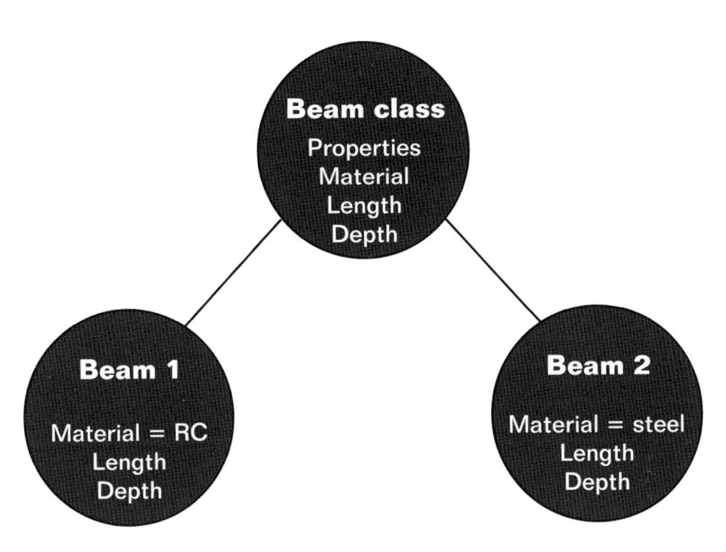

Figure 2.9: Inheritance of class properties (RC = reinforced concrete)

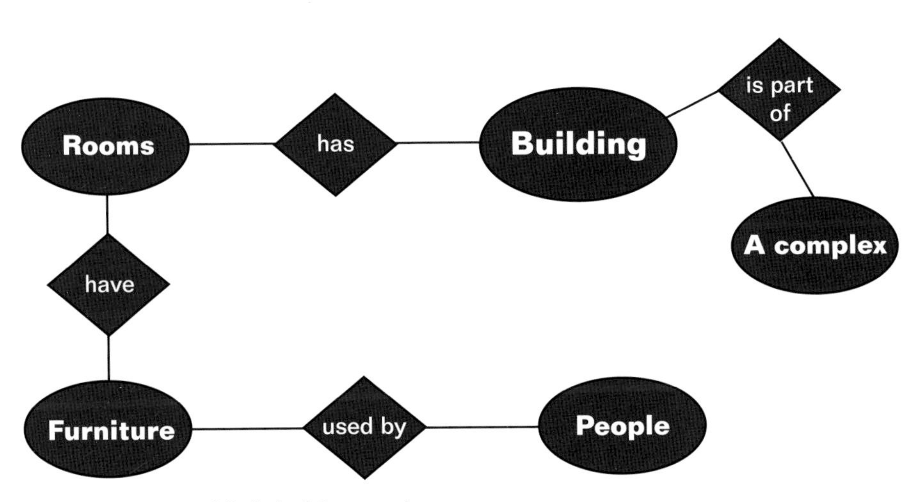

Figure 2.10: E–R model of a building complex

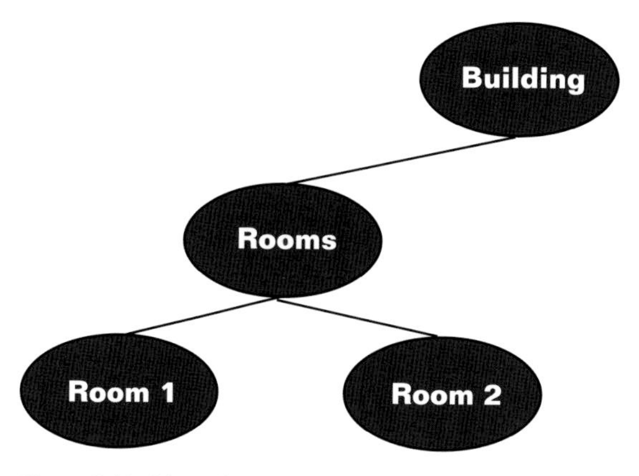

Figure 2.11: Object hierarchies

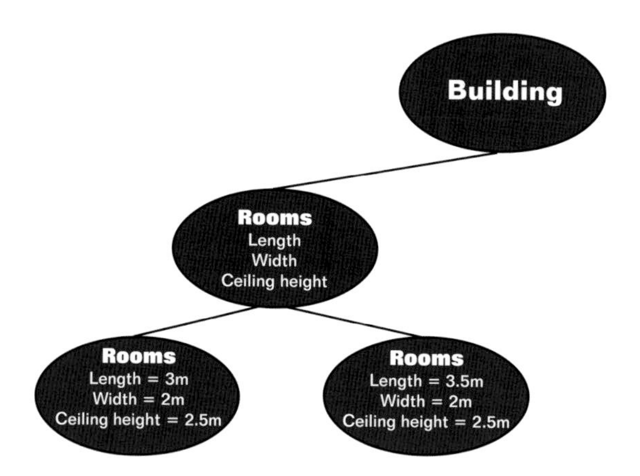

Figure 2.12: Inheritance of properties in object hierarchies

have to be created every time a specific wall is created by a user. It is one of the key and powerful features of an object-based parametric modelling system.

Now, let us consider the importance and use of creating relationships between objects (entities). The E–R diagram in figure 2.10 is a model of a building complex. Although it is a trivial example, it does convey the power behind this modelling technique. The relationship names appear in the diamond-shaped boxes joining two entities.

Using the model based on the E–R diagram shown in figure 2.10, one can query the model for the following:

- Who uses the building?
- Where do you find furniture?
- What is a building made up of?
- etc.

Another example is shown in figure 2.11.

Using the model based on the E–R diagrams shown in figures 2.10–2.12 and the class and object properties shown, one can now also query the model for the following:

- How many rooms are there in the building?
- What is the area of each room and hence the total area of the building?
- etc.

This is exactly how a BIM system generates the different schedules (e.g. room schedules) that are so useful in design and several other activities.

2.8 Definition of BIM

Now that sufficient background material has been provided on relevant modelling and other approaches that BIM is based on, this section provides a definition of BIM. One of the simplest ways to define and understand an acronym is to first break it down into its constituent parts. Therefore, BIM can be broken down as shown below and a meaning or definition provided for each term:

Building is a constructed facility that satisfies the requirements of a dwelling/office/commercial enterprise/etc. However, as mentioned previously, this is a noun version of the word. The verb version of the word is probably more apt in the present context.

As discussed previously, *Information* is a combination of raw data that conveys a meaningful message.

Modelling is an act of describing/representing anything by any means to develop further understanding.

There are several definitions of BIM provided by different people and a selection of these is shown below.

BIM is a *modelling technology* and *associated set of processes* to produce, communicate and analyse building models. These building models are characterised by (Eastman *et al.*, 2012):

- Building components that are represented with intelligent digital representations that 'know' what they are and can be associated with computable graphic and data attributes and parametric rules.
- Components that include data that describe how they behave, as needed for analyses and work processes, e.g. take-off, specification and energy analysis.
- Consistent and non-redundant data such that changes to component data are represented in all views of the component.
- Co-ordinated data such that all views of a model are represented in a co-ordinated way.

Paul Morrell (ex-Chief Construction advisor to the UK Government) believes that BIM is about 'the intelligent use of digital data to design, construct, manage and use a built facility' (as quoted in *AEC Magazine*, 2012).

The Royal Institute of Chartered Surveyors (RICS) describes the fundamentals of BIM as '... a common single and co-ordinated source of structured information...' (www.rics.org/uk).

The National BIM Standards – United States definition is 'A BIM is a digital representation of physical and functional characteristics of a facility. As such it serves as a shared knowledge resource for information about a facility forming a reliable basis for decisions during its life cycle from inception onward' (NBIMS, 2013).

A slightly different way of looking at BIM, particularly in relation to CAD, is that BIM contains *syntax* (geometry, topology) as well as *semantics* (meaning). On the other hand, CAD contains only *syntax*. Therefore, *semantics* is what gives BIM the additional power and strength to be able to facilitate things that CAD cannot. Semantics, in this context, is simply the meaning provided by the additional information attached to every element in a building information model.

2.9 What is NOT BIM technology?

Just as it is important to understand what BIM is, it is also important to appreciate what is *not* BIM. The reason for this importance is that simply by looking at a visual of a building information model or a CAD 3D (three-dimensional) model, it is virtually impossible to differentiate between them. Here are some key pointers to look for in relation to what is *not* BIM (Eastman *et al.*, 2012):

- models that contain 3D data only without any object properties
- models that cannot support behaviour rules for objects
- 2D (two-dimensional) models that must be aggregated together to define a whole building
- models that cannot propagate changes in one view automatically to other views

2.10 Differences between CAD and BIM technologies

Continuing on the same theme, here are some more pointers to particular differences between CAD and BIM:

- CAD provides a 'dumb' drawing, a collection of various geometric shapes and components that make up a drawing of a building.
- BIM provides an 'all-encompassing' container of information about all aspects of a building, i.e. a building information model should contain information about design, construction, operation, schedules, costs, etc., for a building. Therefore, it can be used to extract information and reason with it for all imaginable aspects of a building!

In CAD the core entity is a drawing. However, in BIM the core entity is building objects with attached information as parameters as well as rules about their behaviour. BIM is based on *object-based parametric modelling* as briefly described previously. This is the key difference between BIM and CAD.

2.10.1 Geometric and building data models

Figure 2.13 illustrates the difference between a traditional geometric data model as contained in a 2D CAD drawing and a building model typically found in building information models.

The main difference is the ability to 'model' concepts such as *space* in addition to physical objects (Khemlani, 2004). This requires the ability to include information regarding topology (i.e. connectivity between different elements of a building). For example, a

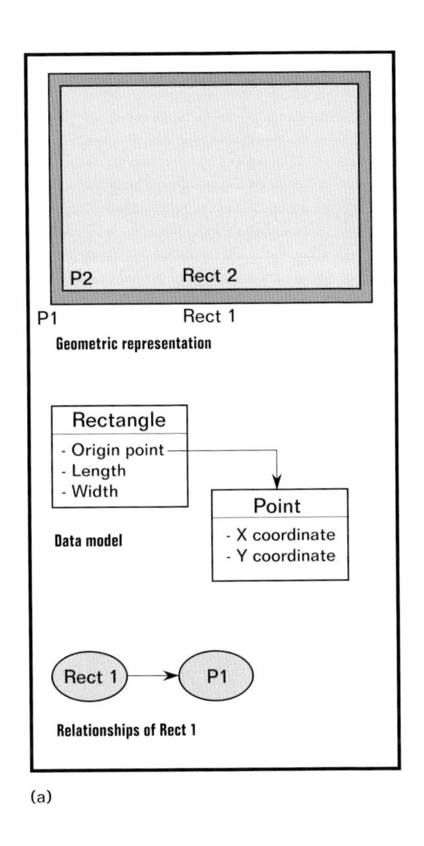

(a)

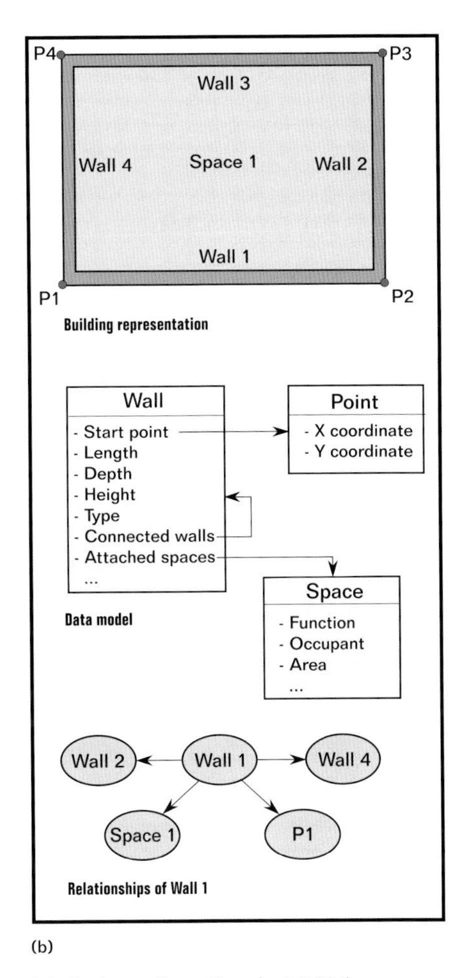

(b)

Figure 2.13: (a) Geometric and (b) building data models (redrawn from Khemlani, 2004) *wall* may enclose a *space* and may be connected to other *walls* as shown in figure 2.13. In addition, even a concept such as *space* can be dealt with and reasoned with by storing and manipulating various *parameters* that define its major characteristics such as its function (office space, circulation space, etc.), its area, its occupants, and so on as shown in figure 2.13(b). This is exactly what a BIM system would do to model *space*.

2.10.2 GUIs of CAD and BIM tools

One of the simplest ways of identifying the difference between a CAD and a BIM tool is by examining the GUIs of typical CAD and BIM systems.The CAD system deals with primitive objects such as lines, curves, circles, etc., whereas the BIM tool deals with primitives such as doors, windows, columns, roofs, etc.

If one thinks about this, it has major implications not for the way the two tools work but also for the end users of these systems. A CAD system user does not necessarily need to have any understanding of how a building is put together (although that would help), but it is practically impossible for anyone to use a BIM system without at least having some understanding of how buildings are put together. It is, therefore, quite clear that the BIM tools are fundamentally different from CAD systems in the way that they work and how users need to use them. To further elaborate on this, figure 2.14 is from a popular BIM tool, with the purpose being to illustrate the components of a

Figure 2.14: GUI of a BIM system (source: Autodesk Inc.)

typical BIM tool user interface. In addition to the *ribbon* across the top of the screen, there are typically three parts of the screen underneath it. On the left, there are two smaller windows, whilst a larger window is positioned on the right, which displays a model of a building. As the intention here is not to give an elaborate introduction to a BIM GUI, the window entitled *Properties* only will be discussed, as it is key to the BIM way of doing things in this context. This window is where all the parameters of all the components/elements of the building are displayed. Depending on which element is selected by the cursor in the right window, all its properties are displayed in the *Properties* window.

By clicking on the end wall of the model shown in figure 2.14, the *Properties* window on the left will show all the different parameters and their values for the wall in question, such as dimensions, area, volume, etc. This set of properties can be customised by adding or modifying the default set of properties that a typical wall comes with in a BIM tool. By having the ability to store key parameters and the ability to manipulate them as required makes these tools immensely more useful and powerful than a typical CAD tool. Because of these stored properties with associated values, it is now possible to extract all kinds of information and data directly from the model that would normally require moving between several systems if one were to use a CAD system, making the whole process cumbersome and prone to errors.

Therefore, another way of looking at the transition from CAD to BIM tools is to say that CAD tools deal with *data*, whereas BIM tools deal with *information*. Figure 2.15 conveys this point graphically. The next transition would be to move on to the next key point in the data-information-knowledge spectrum, i.e. when tools can process *knowledge* about buildings. At that point, perhaps one would be dealing with a new set of BKM (Building Knowledge Modelling) tools. There are already some examples of research prototypes addressing some of the BKM issues (Fruchter *et al.*, 2009; Voss *et al*, 2012). However, there is some way to go yet before these efforts become a commercial reality.

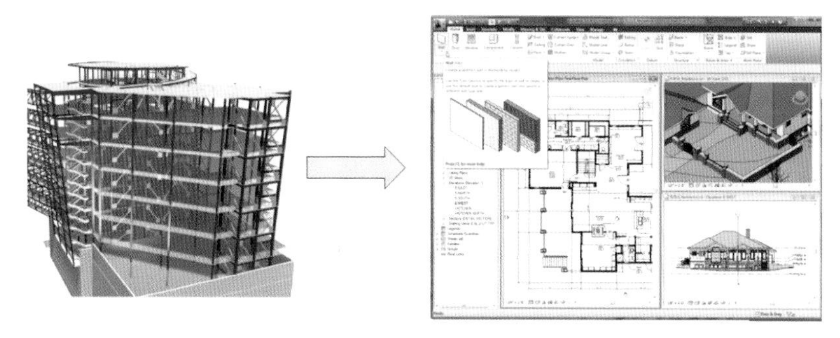

Figure 2.15: Transition from CAD to BIM

2.11 Common uses and benefits of BIM tools

This section provides some common uses of BIM technologies. One of the earliest examples of using this technology was to identify clashes between building elements. Figure 2.16 shows one such clash between a beam and the lift shaft detected easily and quickly picked up by BIM software. One associated benefit with significant impact on productivity concerns the way these clashes can be fixed using a BIM system. Of course, one of the key ideas behind using BIM tools is to *build* the asset twice, once *digitally* and fix all these issues of constructability, etc., before *building* it *physically* on site. But, quite importantly, before BIM came along, the modification to one element of the building generally meant modifying several other drawings too, a dreadful scenario for the designers and draughtsmen! BIM does all the modifications to all the dependent elements automatically, thus, saving an enormous amount of time. Thinking about how this happens, it is not really that difficult when one considers the technology behind it. It does this by using the relationships in the E–R approach to information modelling discussed previously.

Another early and useful example of the usage of BIM tools is the impact they have on ensuring that all the dimensions and labelling are consistent in a model. Again, they do this using the same principles of relationships and other characteristics of object-based parametric modelling.

A similar initial application of BIM technology was co-ordination between different drawings by different stakeholders, such as architects and structural engineers.

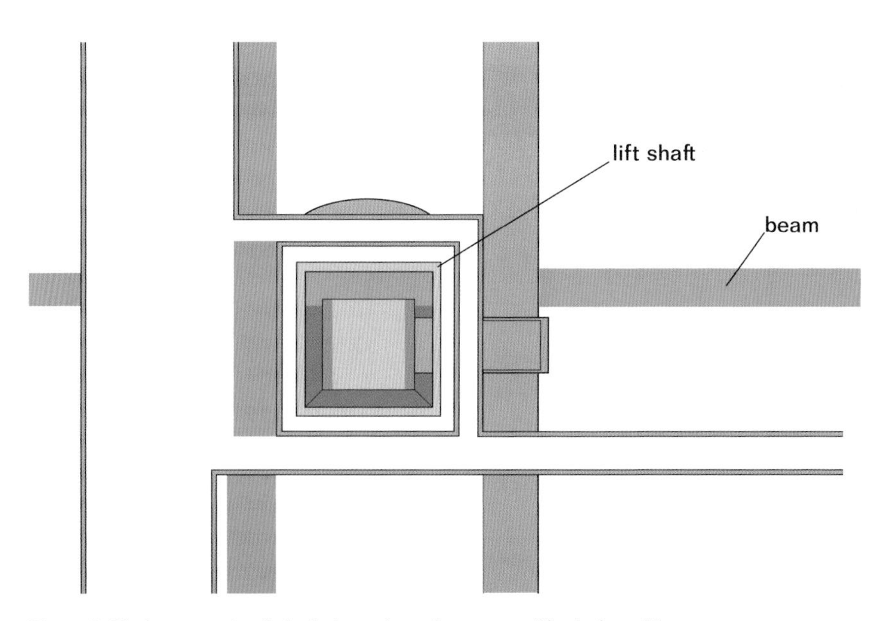

Figure 2.16: An example of clash detection – between a lift shaft and beam.

Here is a list of the most commonly used applications of BIM technologies:

- design visualisation
- material and cost estimation
- quantity take-offs
- clash detection
- consistent drawings
- incorporation of additional *information* and *knowledge* into the model itself, which could be used by other disciplines downstream

More recently, some organisations have also started using BIM for several other purposes. Here is a selection:

- condition monitoring using 3D laser scanning
- integrated design and fabrication
- design checking and assessment
- integrated energy performance analysis
- integrated analysis of various issues such as way-finding or crowd behaviour

2.12 Four-dimensional modelling

Conceptually, 4D (four-dimensional) CAD is a medium representing time and space. It is a type of graphic simulation of a construction schedule. In construction, a 4D animation simulates the process of transforming space over time, thereby giving a clear sense of how a project might be progressing over time in a graphical way.

Producing a 4D animation essentially involves linking a 3D graphic model and a construction schedule. The result of this linking process is a 4D model that represents the physical model of the constructed facility.

2.12.1 Is 4D model a Building Information Model?

'Yes and no' is the short answer! The key criterion for BIM is the underlying parametric model of the asset. If the fourth dimension of schedule is added on top of a parametric model, then a 4D model is a building information model. However, if the fourth dimension of schedule is added on top of a 3D solid CAD model of an asset, then the answer is a straightforward 'no'! Interestingly, both examples are frequently referred to as examples of 4D models in the wider industry, and there are also examples of both approaches commercially available on the market.

2.12.2 Benefits of BIM

Figure 2.17 (Bew, 2010) lists some of the key benefits of BIM by linking them to the various stages of the older RIBA (Royal Institute of British Architects) PoW (Plan of

What is Building Information Modelling?

Building Information Modelling (BIM) is the process of generating and managing information about a building during its entire life cycle.

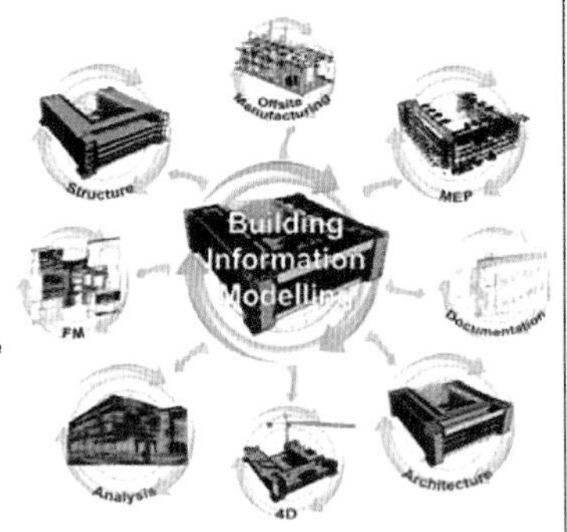

BIM is a suite of technologies and processes that integrate to form the 'system' at the heart of which is a component-based 3D representation of each building element; this supersedes traditional design tools currently in use. Each component is generated from a product library and has embedded information about the product and its placement, material, specification, fire rating, U-value, fittings, finishes, costs, 'carbon content' and any special requirements, which is stored in the system. As the design progresses, so the integrated information becomes more valuable. Sophisticated applications and clash detection can rapidly identify issues which can be designed out at an early stage. BIM could be mistaken for a simple design tool but this overlooks the fact that it is the way the system generates interfaces to and uses information from other systems which is fundamental to the delivery of greater benefits. These benefits accrue to the whole supply chain through the collaborative, integrated use of BIM.

There are parallels between BIM and the EPOS (electronic point of sale) and ERP (enterprise resource planning) systems ubiquitously found in the retail sector. There is a clear interface between BIM and organisational corporate systems – including those dealing with procurement, finance and supply chain performance. As the graphic (above right) shows, BIM sits at the heart of an information web which includes systems directly attributed to design, construction and those business support systems which are used to ensure overall efficient asset and corporate performance.

Why Should I Consider BIM?

There is some published commercial research and data gathered from case studies. The case studies mentioned as 'key projects' in the right-hand panel on the previous page have been summarised in the graph (right).

The red plot indicates the average saving of projects available for measurement to date, with the green plot indicating potential benefits expected by the early adopter community.

Most of the data covered the design, pre-construction and construction stages. There were clear wins in design understanding, spatial and design co-ordination and 4D programme integration. Benefits post construction have still to be measured. Clearly, this demonstrates considerable upside potential, especially when the processes are applied to the entire project life cycle through to facilities management. These benefits increase further year on year, as around 80% of an asset's cost is incurred during its operational phase. Whilst our case studies have focused on monetary benefits, studies are ongoing to establish similar relationships for carbon.

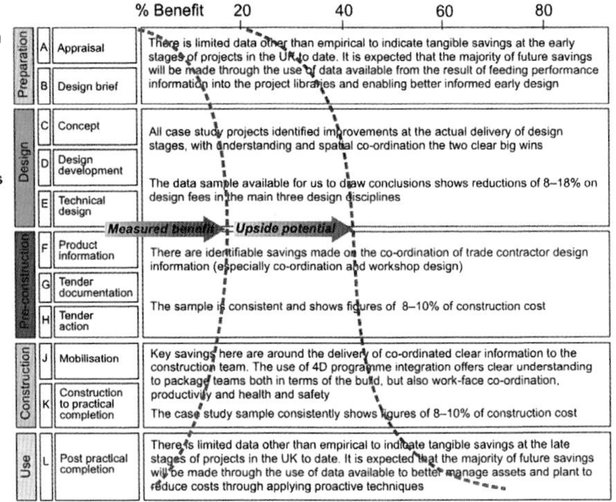

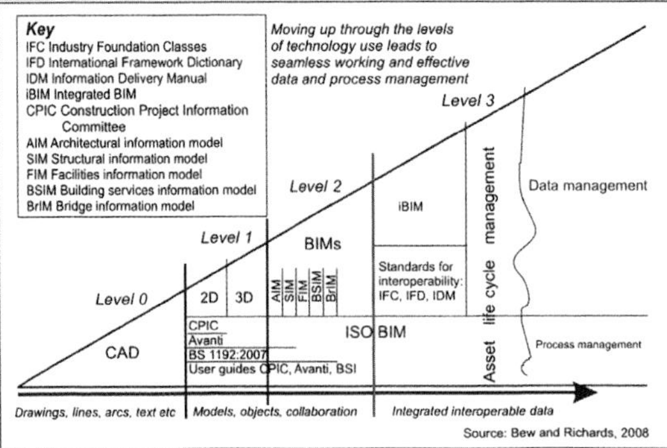

Evolution Not Revolution

As this illustration shows, BIM continues to develop. Clearly, not all businesses will adopt systems and technologies at the same rate. However, just like organisations in the retail sector before them, BIM adopters will need to go though a managed process of change which encompasses not only their internal organisation but also the way they interface with their external supply-base and clients. The majority of the UK market is still working with Level 1 processes, and the best in class are experiencing significant benefits by moving to Level 2. It is clear that organisations adopting BIM now will be those most likely to capitalise on this advantage as the market improves.

Figure 2.17: Benefits of BIM (reproduced from Bew, 2010)

Works). Clearly, BIM has something to offer across all phases of design, construction and O&M phases of asset procurement. However, one should treat this with caution in that the real benefits of BIM can only accrue when the whole project team uses standard processes and protocols for exchanging information at key stages. This requires a completely different way of working and procuring assets. As mentioned earlier, this cannot happen with technologies alone and, above all, without a change in culture and mindset of the industry.

2.13 Summary

This chapter has presented basic concepts of information modelling and BIM. Some common and popular uses of BIM technologies have also been outlined. However, it should be pointed out that all the examples of BIM uses provided in this chapter relate to the use of technology almost in complete isolation of processes, standards and protocols for information exchange that are so vitally important for obtaining the real benefits of BIM in a project. These uses are sometimes referred to as *lonely* BIM uses. It is important to appreciate that although an organisation can gain significant benefits internally through the use of BIM technologies, unless associated processes, standards and protocols are in place for the whole project team, the real benefits to the project and hence the client are not going to accrue. This point cannot be overemphasized, it is so important.

3

UK Government's BIM strategy

3.1 Introduction

In 2011 the UK Government published its comprehensive Construction Strategy (2011). Section 2.29 of the strategy states that 'at the industry's leading edge, there are companies which have the capability of working in a fully collaborative 3D environment, so that all of those involved in a project are working on a shared platform with reduced transaction costs and less opportunity for error; but construction has generally lagged behind other industries in the adoption of the full potential offered by digital technology'. The section is actually entitled BIM specifically as the way forward for achieving 20% efficiency savings in the industry. The UK Government has a simple hypothesis 'Government as a client can derive significant improvement in costs, value and carbon performance through the use of open shareable asset information'. Therefore, the UK Government mandated the use of Level 2 as a minimum on all publically procured projects from 2016. The ultimate goal for the Government (similar to that of any other client organisation) is to save money in O&M and to reduce the carbon footprint of the assets it owns. It is well established that more than 60% of the life-cycle costs of constructing and maintaining a building is spent beyond the handover phase at the end of the construction period. Clearly, any savings that the Government can make as a huge asset owner can ultimately result in very substantial savings to the taxpayer. This chapter gives an overview of the UK Government's BIM strategy and the main drivers behind it. It also provides an introduction to some other related and linked issues that form the core elements behind the strategy's implementation that includes the GSL (Government Soft Landings), 'shrink-wrapped' guidance documents published by the BIM Task Group and RIBA's latest Plan of Work 2013.

3.2 Main drivers and goals

As mentioned earlier, the main drivers behind its BIM strategy for the Government are saving costs and reducing the carbon footprints of its asset base. More recently, the Government has published the Construction 2025 (2013) document detailing its vision and expectations from this industry by 2025. Figure 3.1 summarises the kinds of results that the Government expects to achieve by working together with the industry such that by 2025 the construction industry will:

- be known for the diverse and talented *people* in its workforce

- be *smart* and efficient through the use of state-of-the-art technologies
- lead the world in low-carbon and *sustainable* development
- drive *growth* across all sectors of the economy
- have a clear *leadership* from a Construction Leadership Council

On first glance, the numbers appear to be quite ambitious, but if one breaks them down into where these results will come from, then it becomes clear that although not easy, achievement does appear possible. It should be pointed out that this report is not a report on BIM strategy, but that all the targets mentioned here do appear to be proposed on the premise that the industry would be BIM-driven from inception to O&M.

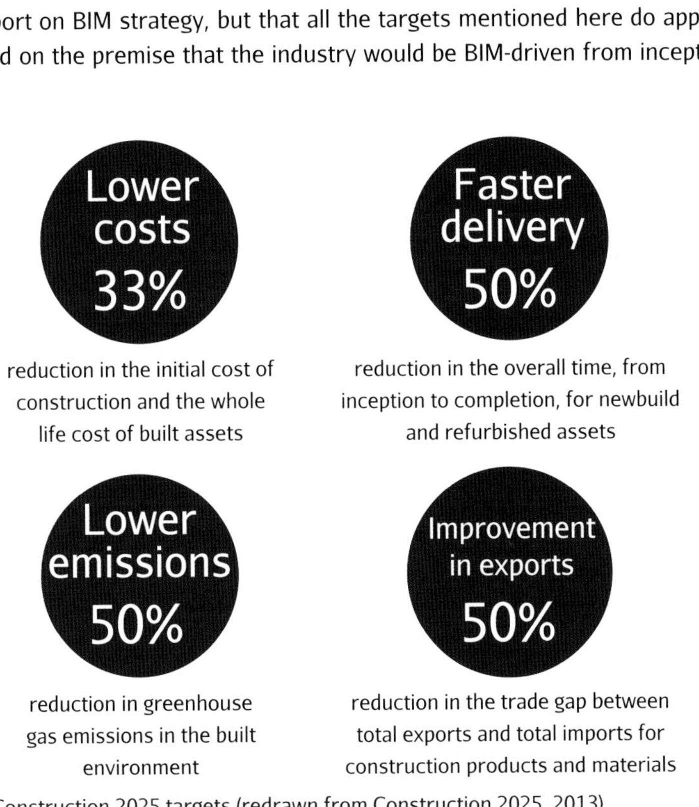

Figure 3.1: Construction 2025 targets (redrawn from Construction 2025, 2013)

In order to promote the use of BIM processes and technologies, the Government set up a BIM Task Group, which was charged with developing guidance and implementing various other activities to promote BIM within the construction industry. The Task Group began its work in 2011 by first setting up a number of subgroups charged with dealing with specific aspects of BIM adoption by the industry, e.g. developing standards for processes, protocols and other related issues such as guidance on training and education. Right at the outset, drawing upon an earlier study (Bew and Richards, 2008), the Task Group laid down very clear guidance on what it meant by BIM adoption by specifying Level 2 BIM as a minimum standard to be achieved on all publicly funded projects by 2016. Level 2 is one of four maturity levels of computer-based design technologies and associated processes, which are defined in section 3.3. The ultimate level of BIM adoption in projects will be the next level, Level 3, which will require a fully shareable and collaboratively owned model that the Government is hoping to mandate in due course a few years hence. Therefore, Level 2 BIM by 2016 is seen as an important staging milestone on the road to Level 3 at some point in the future. This is a pragmatic decision because Level 3 involves complex legal and contractual issues that may not be easily resolvable in the foreseeable future, e.g. shared ownership of models implies shared liabilities, which is a difficult concept to handle in an appropriate legal and contractual framework.

This chapter gives an overview of the key elements of the UK Government's BIM strategy and discusses how they lay down a solid foundation for the industry to get up to speed by 2016. The key specific elements will be dealt with in greater detail in subsequent chapters.

3.3 Maturity levels

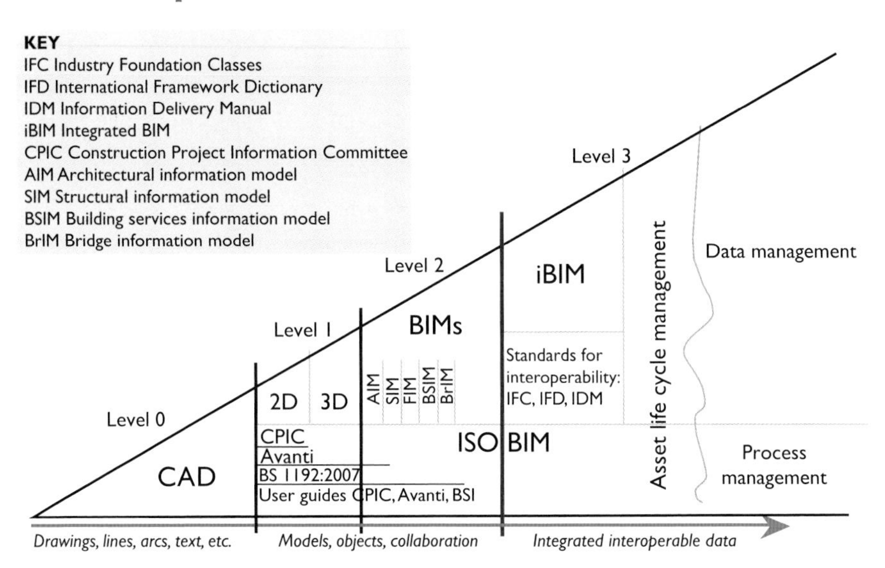

Figure 3.2: 'Wedge diagram' (redrawn from Bew and Richards, 2008)

Level 0 – Unmanaged CAD, probably 2D, with paper (or electronic files) as the most likely data exchange mechanism.

Level 1 – Managed CAD in 2D or 3D format using BS 1192:2007 (BSI, 2008) with a collaboration tool providing a CDE (Common Data Environment) and possibly some standard data structures and formats. Commercial data managed by stand-alone finance and cost management packages with no integration.

Level 2 – Managed 3D environment held in separate discipline 'BIM' tools with attached data. Commercial data managed by ERP (enterprise resource planning) systems. Integration on the basis of proprietary interfaces or bespoke middleware could be regarded as 'pBIM' (Proprietary BIM). The approach may utilise 4D programme data and 5D (five-dimensional) cost elements.

Level 3 – Fully open process and data integration enabled by IFC (Industry Foundation Class) / international framework dictionary. Managed by a collaborative model server. Could be regarded as iBIM (Integrated BIM), potentially employing concurrent engineering processes.

3.3.1 Level 2 – key issues

Key to the strategy is the need to deliver Level 2 BIM capabilities with combined model, drawing and COBie (Construction Operations Building Information Exchange) data deliveries to the client at key points throughout the delivery and handover processes.

To achieve this, clear contractual and delivery guidance is being made available to the supply chain. PAS 1192:Part 2 and Part 3 (CIC, 2013b; CIC, 2014) form the first set of guidance documents. On successful completion of the early adopter programmes, the documents and processes will be refined and converted into full British Standards.

The central idea behind Level 2 BIM is that everyone uses the same *data formats* and *timings for data delivery* and the data format to be used will be COBie. Level 2 BIM essentially advocates for a federated collaboration model, as shown in figure 3.3.

This essentially means that the ownership of models (architectural, structural, MEP (Mechanical, Electrical and Plumbing), etc.) stays with the creators of the models, but they share the models through a central CDE (Common Data Environment), as shown in figure 3.3. In Level 2 BIM, the ownership (and hence the liabilities) are not shared, thus keeping things much more manageable. From an implementation point of view, the major change being mandated is simply that all stakeholders move on to using BIM (rather than 2D/3D CAD) technologies and comply with information standards and protocols for exchange at well-defined *data drop* points along the life cycle of an asset from concept inception to demolition (and perhaps rebuild). The standards for information representation, storage and exchange will be agreed right at the beginning of a project and will be contractually binding on all stakeholders. This is the key change. An important implication of this is that moving on to Level 2 BIM has very little (if any) impact on existing contracts used by the industry. The BIM Task Group has clearly taken this

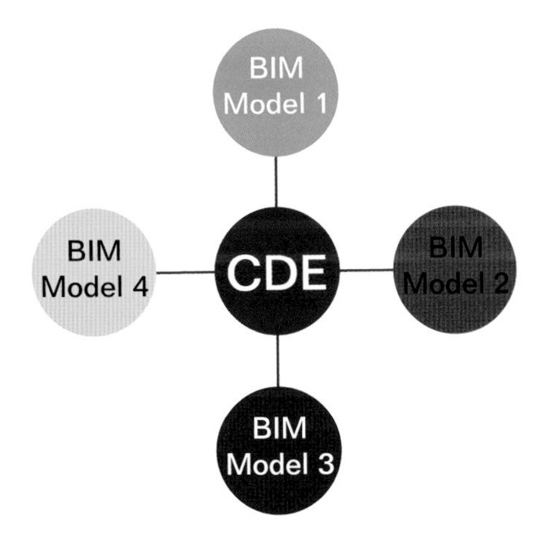

Figure 3.3: Level 2 federated BIM architecture

approach as the only viable and pragmatic solution to encouraging an industry to *change to a new way of working*. The transition to Level 2 BIM essentially is a new way of working and delivering assets in this industry. The idea is that the processes and deliverables that will be mandatory will ensure the contribution and recognition of all stakeholders right from the start of any project, thereby minimising any potential changes (and hence increased costs and schedules) downstream. It is also the idea that every asset is *built twice, once digitally and once physically*, after all issues have been digitally resolved by all stakeholders working collaboratively from a very early phase of the life cycle.

3.4 GSL

The Government's Construction Strategy (2011) pointed out the discrepancies between the built asset's actual required performances, and suggested that *soft landings* was one of the ways to address this issue. In light of this, a GSL was developed. It is a matter of common knowledge that the cost of maintaining and operating a building is considerably more than the capital costs of design and construction of the building. Typically, O&M costs over the lifetime of a building are in excess of 60% of the overall life-cycle costs. Therefore, in order to make savings throughout the life cycle of an asset, there is a need to manage the O&M costs more carefully and ensure that everything possible is being done by the O&M team to optimise these costs. The traditional view was to focus on O&M costs in isolation of the earlier phases of the life cycle, design and construction. However, it is now realised that O&M costs can only be properly optimised if the key factors that consume resources in this phase are given due cognisance right from the inception and design stages. Therefore, GSL recognises this quite explicitly and spells out the need for early recognition of O&M costs at the design phase.

GSL was originally developed by the Government Property Unit in 2011 in partnership with a number of industry experts. Latterly, GSL has been aligned with the stages and processes put forward by the BIM Task Group. The GSL Implementation Guide, published originally to provide guidance on how to embed GSL in delivering assets, has been brought under the remit of the Task Group. In September 2012, the GSL policy was agreed by the Government Construction Board and is mandated to be implemented on all public sector projects, including refurbishments, by 2016, thereby aligning it with the Government's BIM Level 2 policy. Note the road map for 2016 BIM Level 2 shown in figure 3.4, which demonstrates the alignment with GSL.

During the course of 2013, central government departments were asked to implement GSL in their projects. Every department was supposed to identify a GSL champion, and a GSL Stewardship Group was set up to take on board any lessons learnt to feed back into the Implementation Guide with a view to having this ready in time for the 2016 mandated implementation on all projects.

As stated in the Government's Construction Strategy (2011), the main objective of GSL is *aligning the interests of those who design and construct an asset with those who subsequently use it*.

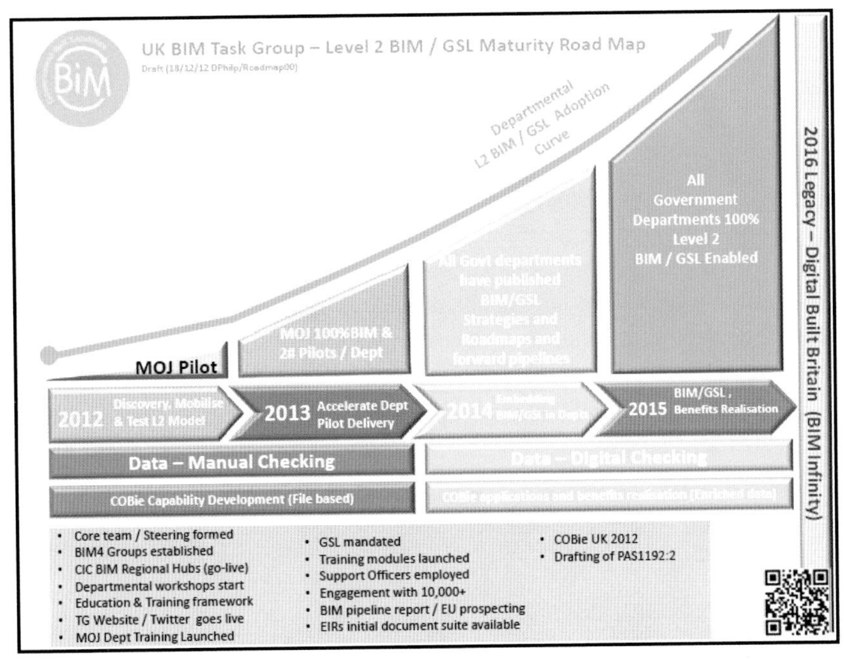

Figure 3.4: BIM Task Group's road map for 2016 (reproduced from BIM Task Group website, www.bimtaskgroup.org)

3.5 'Shrink-wrapped' guidance documents

The UK BIM Task Group has published a number of documents that are meant to help the industry get up to speed by 2016. These include the following:

BIM protocol. A universal addendum to appointment documents and construction contracts that puts in place additional rights and obligations associated with building information models – includes a guidance note and a model contract amendment.

Employer's Information Requirements (EIR). A template setting out the information needed to define information and other client requirements for a project delivered using building information models – includes guidance and pre-qualification documentation for consultants and contractors.

Outline Scope of Services for the Role of Information Management. A Scope of Services for information management activities that is designed to be appended to an existing appointment, e.g. a design team leader. Information management is referenced in the BIM protocol and promotes the adoption of the CDE described in PAS 1192: Part 2. Versions have been prepared for use with stand-alone schedules of service and the CIC (Construction Industry Council) Schedules of Service.

PAS 1192:Part 2:2013. BIM – production information requirements for capital projects. The standard for information exchange to meet BIM maturity Level 2 – includes the BIM implementation plan and details of the CDE. The PAS is published by the British Standards Institution.

Best Practice Guide for Professional Indemnity Insurance when using Building Information Models – This guidance document provides advice for the insured parties like design consultants as to what they should insure in terms of their professional indemnity cover vis-à-vis production of building information models for their clients. The key message from this document is that there are no issues with level 2 BIM warranting any coverage restrictions for those companies or individuals and therefore any impact on insurance premium will be minimal, if any.

PAS 1192:Part 3:2014 Specification for information management for the operational phase of assets using building information modelling – This guidance is a direct follow on from part 2 outlined above and deals exclusively with the post-handover phase of an asset delivery. The main issues pointed out here are that the transient and often unprecdiatable nature of operation and maintenance (O&M) beyond handover warrants a different non-sequential process flow as opposed to the pre-handover which are largely iterative but sequential in nature. This document has been published very recently and therefore any comprehensive treatment of this publication is outside the scope of this book.

BS 1192:Part 4:2014 Collaborative Production of Information – This publication exclusively deals with COBie-based information production and exchange.

The purposes of these documents include the following:

- acceleration of the adoption of Level 2 BIM with simple, standard documents
- alignment with existing contracts and appointments
- clear communication of requirements for data
- address potential blockers such as PII
- formalise data and information management practice

In addition, three further documents are in various stages of development and release. Readers are encouraged to keep a watch on all the relevant websites for developments:

- PAS 1192–2015: Specification for security-minded building information modelling, digital built environments and smart asset management (recently released)
- DPoW (Digital Plan of Works) (released on the beta version in April 2015)
- classification (Uniclass 2015 which has some tables released and others in beta versions)

3.6 RIBA PoW 2013

In this section, the latest version of RIBA's PoW (figure 3.5) is briefly introduced as it is a key part of delivery guidance for any project in the absence of a DPoW. This new version has been developed particularly in view of the BIM strategy of the Government and the direction being set by these developments for the industry. The key difference

Tasks ▼ / Stages ▶	0 Strategic Definition	1 Preparation and Brief	2 Concept Design	3 Developed Design	4 Technical Design	5 Construction	6 Handover and Close Out	7 In Use
Core Objectives	Identify client's **Business Case** and **Strategic Brief** and other core project requirements.	Develop **Project Objectives**, including **Quality Objectives** and **Project Outcomes**, **Sustainability Aspirations**, **Project Budget**, other parameters or constraints and develop **Initial Project Brief**. Undertake **Feasibility Studies** and review of **Site Information**.	Prepare **Concept Design**, including outline proposals for structural design, building services systems, outline specifications and preliminary **Cost Information** along with relevant **Project Strategies** in accordance with **Design Programme**. Agree alterations to brief and issue **Final Project Brief**.	Prepare **Developed Design**, including coordinated and updated proposals for structural design, building services systems, outline specifications, **Cost Information** and **Project Strategies** in accordance with **Design Programme**.	Prepare **Technical Design** in accordance with **Design Responsibility Matrix** and **Project Strategies** to include all architectural, structural and building services information, specialist subcontractor design and specifications, in accordance with **Design Programme**.	Offsite manufacturing and onsite **Construction** in accordance with **Construction Programme** and resolution of **Design Queries** from site as they arise.	Handover of building and conclusion of **Building Contract**.	Undertake **In Use** services in accordance with **Schedule of Services**.
Procurement *Variable task bar	Initial considerations for assembling the project team.	Prepare **Project Roles Table** and **Contractual Tree** and continue assembling the project team.	←- - - - The procurement strategy does not fundamentally alter the progression of the design or the level of detail prepared at a given stage. However, **Information Exchanges** will vary depending on the selected procurement route and **Building Contract**. A bespoke **RIBA Plan of Work 2013** will set out the specific tendering and procurement activities that will occur at each stage in relation to the chosen procurement route. - - - -→			Administration of **Building Contract**, including regular site inspections and review of progress.	Conclude administration of **Building Contract**.	
Programme *Variable task bar	Establish **Project Programme**.	Review **Project Programme**.	Review **Project Programme**.	←- - The procurement route may dictate the **Project Programme** and may result in certain stages overlapping or being undertaken concurrently. A bespoke **RIBA Plan of Work 2013** will clarify the stage overlaps. The **Project Programme** will set out the specific stage dates and detailed programme durations. - -→				
(Town) Planning *Variable task bar	Pre-application discussions.	Pre-application discussions.	←- - - - - - Planning applications are typically made using the Stage 3 output. A bespoke **RIBA Plan of Work 2013** will identify when the planning application is to be made. - - - - - -→					
Suggested Key Support Tasks	Review **Feedback** from previous projects.	Prepare **Handover Strategy** and **Risk Assessments**. Agree **Schedule of Services**, **Design Responsibility Matrix** and **Information Exchanges** and prepare **Project Execution Plan** including **Technology** and **Communication Strategies** and consideration of **Common Standards** to be used.	Prepare **Sustainability Strategy**, **Maintenance and Operational Strategy** and review **Handover Strategy** and **Risk Assessments**. Undertake third party consultations as required and any **Research and Development** aspects. Review and update **Project Execution Plan**. Consider **Construction Strategy**, including offsite fabrication, and develop **Health and Safety Strategy**.	Review and update **Sustainability**, **Maintenance and Operational** and **Handover Strategies** and **Risk Assessments**. Undertake third party consultations as required and conclude **Research and Development** aspects. Review and update **Project Execution Plan**, including **Change Control Procedures**. Review and update **Construction** and **Health and Safety Strategies**.	Review and update **Sustainability**, **Maintenance and Operational** and **Handover Strategies** and **Risk Assessments**. Prepare and submit Building Regulations submission and any other third party submissions requiring consent. Review and update **Project Execution Plan**. Review **Construction Strategy**, including sequencing, and update **Health and Safety Strategy**.	Review and update **Sustainability Strategy** and implement **Handover Strategy**, including agreement of information required for commissioning, training, handover, asset management, future monitoring and maintenance and ongoing compilation of **'As-constructed' Information**. Update **Construction** and **Health and Safety Strategies**.	Carry out activities listed in **Handover Strategy** including **Feedback** for use during the future life of the building or on future projects. Updating of **Project Information** as required.	Conclude activities listed in **Handover Strategy** including **Post-occupancy Evaluation**, review of **Project Performance**, **Project Outcomes** and **Research and Development** aspects. Updating of **Project Information**, as required, in response to ongoing client **Feedback** until the end of the building's life.
Sustainability Checkpoints	**Sustainability Checkpoint — 0**	**Sustainability Checkpoint — 1**	**Sustainability Checkpoint — 2**	**Sustainability Checkpoint — 3**	**Sustainability Checkpoint — 4**	**Sustainability Checkpoint — 5**	**Sustainability Checkpoint — 6**	**Sustainability Checkpoint — 7**
Information Exchanges (at stage completion)	**Strategic Brief**.	**Initial Project Brief**.	**Concept Design** including outline structural and building services design, associated **Project Strategies**, preliminary **Cost Information** and **Final Project Brief**.	**Developed Design**, including the coordinated architectural, structural and building services design and updated **Cost Information**.	Completed **Technical Design** of the project.	**'As-constructed' Information**.	Updated **'As-constructed' Information**.	**'As-constructed' Information** updated in response to ongoing client **Feedback** and maintenance or operational developments.
UK Government Information Exchanges	Not required.	Required.	Required.	Required.	Not required.	Not required.	Required.	As required.

***Variable task bar** – in creating a bespoke project or practice specific RIBA Plan of Work 2013 via www.ribaplanofwork.com a specific bar is selected from a number of options.

Figure 3.5: RIBA's PoW 2013 (reproduced from RIBA, 2013)

between earlier PoWs and this one is the fact that for the first time, the core 'actor' in the project has been moved from the architect to the project team. This is significant and is an acknowledgement of how the workflows in the industry and the thinking and approach behind a project execution and delivery have shifted in recent times.

As is well known, there are several other PoWs in operation in the industry (such as CIC work stages, Governance for Railway Investment Projects (GRIP), etc.). Therefore, the following sections give a brief discussion on how the data drops map onto the different stages of some of the more popular and commonly used PoWs.

3.7 Alignment with industry PoWs and data drops

As mentioned earlier, the central idea behind BIM-enabled delivery of assets is that all stakeholders use consistent data formats to deliver information at pre-agreed stages of a project. In order to achieve this, a framework for delivering information needs to be developed that clearly defines the information exchange (data drop) points at the different key stages of a project. This framework needs to outline the key stages of information production and refinement as it gets transformed by traversing through the key stages of the project. The key stages of a project exist in various PoWs used in the industry. The framework for information exchange has to align with the PoW, but the data drop points may not exactly align with each and every stage of such PoW stages. This means that there may be a data drop that aligns with some but not all stages of the PoW being followed in a project. The key point is that all the information that the owner (employer) may require at pre-defined stages must be delivered to them by the relevant parties in the agreed format.

One such framework has been developed by members of the BIM Task Group and the industry. This framework has seven generic stages for information refinement: 1, Brief; 2, Concept; 3, Design Development; 4, Production; 5, Installation; 6, As Constructed; 7, In Use. This framework also provides the basis for common terminology for information exchange to be used across the industry. As will be seen in later chapters, the guidance documents published by the BIM Task Group use these generic stages to articulate the requirements for information exchange through the life cycle of an asset.

A key point to understand about these generic stages is that they deal with information and its refinement without specifying who actually produces the information. At each of these generic stages, the employer is in a position to decide by examining the contents of the information whether the project can move to the next stage. This has also given rise to PLQs (Plain Language Questions) that are meant for use by the employer to decide whether the project can move on. These generic stages are, therefore, a way to ensure consistency in producing information for all stakeholders at different stages and are agnostic of technologies used to produce the information as well as the parties responsible for producing them. Finally, the very important aspect of mapping the seven generic stages mentioned above to the different PoWs (like RIBA or CIC work stages) needs to be done to complete this framework. A detailed discussion is outside the scope of this book. However, it should be noted that such a mapping has been done for several PoWs such as RIBA, Transport for London

Corporate Gateway Approval Process (CGAP), Network Rail (GRIP), OGC (Office of Government Commerce) Gateways, among others.

3.8 DPoW (Task Group document version 7.1)

As outlined above, it is imperative that before BIM Level 2 becomes mandatory in 2016, the industry has all the required guidance and support to make the very important and fundamental transition in processes that this entails. One of the key aspects of this transition is to be able to adapt to a new set of processes that align with the existing processes that the industry is most familiar with. Level 2 BIM is largely about efficient and consistent information exchange between all stakeholders and most importantly about handing over a consistent set of information to the employer at the handover stage that can then seamlessly feed onto their CAFM (Computer-Aided Facilities Management) and asset management systems. Therefore the processes currently used by the industry either need to be adapted for this to happen or a new set of processes needs to be developed that articulates the information management and exchange requirements that align with these existing processes. The importance of this cannot be overemphasized. It is with this motivation that a DPoW was developed by the Task Group, which is likely to be published by the summer of 2015 and was released in beta version in April 2015. Although the final version of the DPoW is still under development, this section provides some of the background thinking that has gone into developing it and is based on the beta version of the document available under the Task Group Labs section of the BIM Task Group's website (www.bimtaskgroup.org/task-group-labs).

According to this document (DPoW, 2013), 'the Digital Plan of Works (DPoW) is the

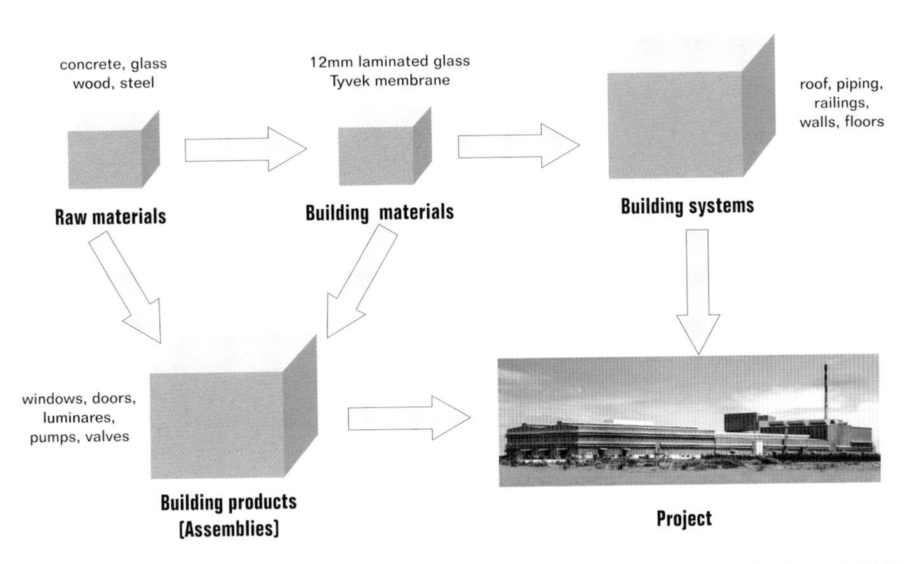

Figure 3.6 A schematic diagram of how these different data types relate to each other (DPoW, 2013).

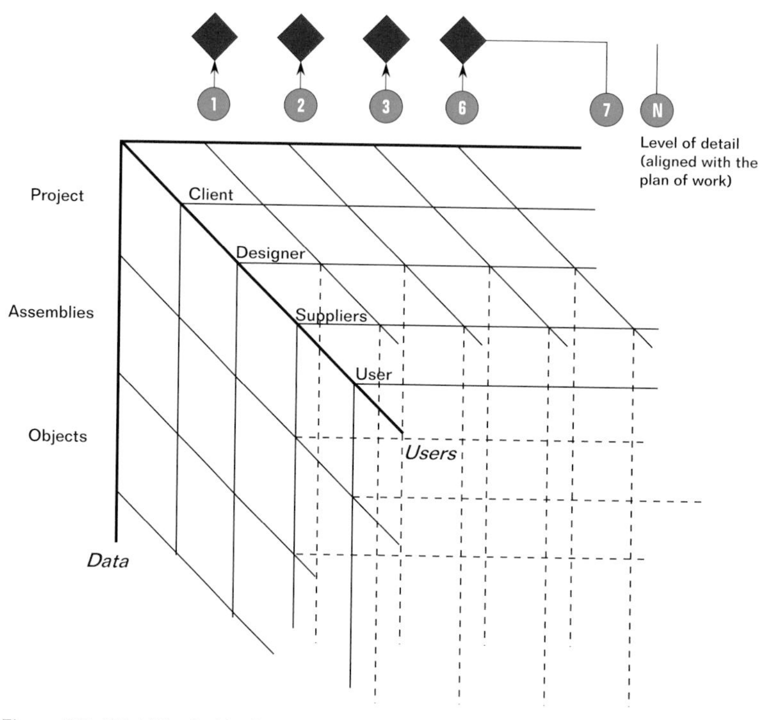

Figure 3.7: DPoW 'cube' (redrawn from DPoW, 2013)

articulation of the project delivery stages and the level of detail/definition that needs to be delivered by each supplier/discipline to the employer at any point in time'. It further adds that this document will ' ... provide a generic framework to enable institutions to overlay local knowledge, taxonomies and brands, whilst preserving consistency in the new underlying data definitions and minimum levels to allow transparent co-ordination across disciplines in multi-discipline projects and schemes'. In fact, the latest RIBA PoW (RIBA, 2013) is a reflection of this philosophy.

3.8.1 Project hierarchy or data types

In order to put things into perspective, it is first important to understand the types of data that a building information model may contain. The relevance of this is that the DPoW is essentially the framework to facilitate the management and delivery of information at defined stages of the life cycle of an asset. Therefore, the first thing to do is to classify the most common types of data that normally constitute any building information model (DPoW, 2013):

- materials
- objects (components)
- assemblies

- projects

The DPoW articulates the relationship between the delivery stages (PoW), geometry and data maturity and the data type hierarchy outlined above. To simplify things, a schematic view of the DPoW is provided as a cube (DPoW, 2013) with the three axes of definitions that need to be navigated, i.e. PoW stages, Level of Definition/Detail and user or audience, as shown in figure 3.7.

A detailed description of this cube is outside the scope of this book. However, it should be clear as to what is required for successfully implementing BIM into projects in terms of the key elements that need to be dealt with. This cube (or a version of this) is also likely to be available on a web portal as the main output of a Technology Strategy Board project with a view to providing guidance to the industry to get up to speed for the Government's target 2016 deadline. The next section will now deal with one of these elements that occupies a key position in the whole process, i.e. LOD (Level of Development).

3.9 LOD and Level of Detail

As explained earlier, the whole point behind BIM-driven project delivery is the development, exchange and storage of information at pre-agreed stages in standard formats. This raises the question of not just *what* but *how much* information should be exchanged during the entire life cycle. In order to obtain more explicit control over the *how much* question, a concept of *LOD* has been developed. A specification for LOD has been provided by various organisations, most prominent among which is the AIA (American Institute of Architects) whose specification is most widely accepted and used. The AIA LOD specification (BIM Forum, 2013) is based on the BIM protocol form 1, AIA G202-2013 (AIA, 2013) and is managed by CSI Uniformat 2010 (Uniformat, 2010). The set of specifications is driven by articulating explicitly how much detail each element in a building information model should contain at different stages. This is clearly of great benefit to model authors and will go a long way in developing consistency of content at the different stages between models developed by different stakeholders of a project. The LOD specification addresses several key issues in relation to information exchange and serves a vital role in facilitating seamless collaboration between all stakeholders in a project. The originators and receivers of information in the shape of models have particular responsibilities to make the communication and collaboration process as seamless as possible. For example, the originators or authors of the models need to specify as to what LOD the elements in the model are at and consequently what the model's limitations might be. With a clear articulation of this, the receivers of the model can then use it appropriately for their purposes. This is also clearly of immense use to the owners of the asset in question, who can use this information to satisfy themselves as to the progress of the project and take necessary steps to ensure that their expectations are being met. Equally, the design managers can use the LOD information to manage their processes at different key stages. Above all, LOD provides a common language for all concerned to ascertain what a model contains and can consequently be used in their deliberations.

It is understood by all that any product passes through several levels of abstraction hierarchy and the design becomes increasingly more specific and detailed as it progresses from the concept to the detailed design stage. The LOD specifications provide a common vocabulary to refer to the design at different abstraction levels for all concerned in a universally agreed standard way. It is a common phenomenon in the industry that designs are often misinterpreted in relation to the original intent of the designers. LOD is a tool to address this important issue.

There is a related concept of *Level of Detail*, which is sometimes confused with LOD. Level of Detail is the *amount of information and detail* in the information content provided at different stages of a project. Generally speaking, Level of Detail is a more detailed, lower level concept, whereas LOD deals with the whole model itself as to what state of development every element in a model may be at a much higher level. LOD deals more with how much design thinking may have been exercised for each element in a model at a point rather than the actual detail of each of these elements. The guidance documents published by the BIM Task Group use both concepts, and a more elaborate discussion on this is deferred to chapter 4.

3.10 Proposed workflow for a typical BIM-enabled project

To summarise the whole process of implementing a BIM-enabled asset life cycle, this section presents a high-level workflow that brings together all the ideas discussed earlier in this chapter. Figure 3.8 shows a high-level view of the workflow that should be typically followed for any BIM-enabled project. The diagram aims to put things into perspective and guide anyone wishing to link the BIM Task Group's 'shrink-wrapped' publications to various stages of a typical project. The process starts with the definition of a need for either a new asset or refurbishment or enhancement of an existing asset. In the case of a new asset, the process starts with a clean sheet with a set of requirements that the asset owner (the employer) might have. This is what would be compiled together in an EIR document. Based on the EIRs, a project procurement process may be initiated, which will be driven by the

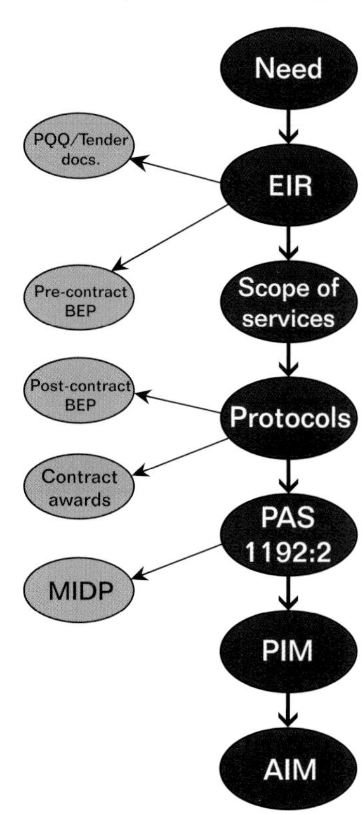

Figure 3.8: Typical workflow for a BIM-enabled asset life cycle

EIR document in terms of which procurement route to adopt as well as informing the tender documents. The tenders could well be single or multistage, but such details are omitted here because the focus is to outline the overall workflow. The tenders received will be based on the EIRs and any protocols that the project may follow. At this stage, the information may well be at a higher level but may include pre-contract BEP (BIM Execution Plan) documents including the tenderers' proposed approach as well as their capability and competence in relation to BIM. This will then be followed by the awards of the contract after various negotiations and clarifications. At this stage, a detailed post-contract BEP document will be prepared, agreed and signed off by all stakeholders of the project. In addition to the BEP document, an MIDP (Master Information Delivery Plan) will be prepared that will be based on and in compliance with the PAS 1192:Part 2 specifications. At this stage, the PIM (Project Information Model) starts taking shape, consisting of essentially graphical (building information models), non-graphical and other documents that will include populated templates from the BEP and other documents as stipulated in the EIR document. At the end of the construction phase, the completed PIM essentially becomes the AIM (Asset Information Model), which is handed over to the asset management and facilities management group. In the case of a refurbishment or enhancement project for an existing asset, the AIM which is already in place for the asset in question becomes the starting point when specifying the need for the project, and which is then followed by the same steps mentioned above for a new build project.

3.11 Summary

This chapter has summarised the UK Government's BIM strategy and provided brief introductions to the various elements of the strategy as well as the core elements of the guidance published by the BIM Task Group for implementing BIM into projects. The following chapters will now deal with each of these elements in detail.

4

EIRs and PAS 1192: Part 2

4.1 Introduction

Earlier chapters have presented the background and main drivers for implementing BIM technologies and processes in construction projects. The UK Government's BIM strategy was presented in chapter 3. Brief overviews of some of the key elements of BIM processes were also provided. This chapter now presents a more detailed discussion on these elements, in particular two of the main 'shrink-wrapped' documents published by the BIM Task Group: the EIR and PAS 1192:Part 2. The purpose here is not a regurgitation of the materials in these documents but to provide guidance on using these documents, concentrating only on those parts that are most and directly relevant to BIM implementation issues in projects.

4.2 EIRs

EIRs set out important guidance to drive the project procurement and delivery processes. The idea is to spell out in so many words and as unambiguously as possible the key sets of data and information as well as the points along the project stages when the employer (client) organisation would require them. Some key facts about EIRs include the following (CIC, 2013a):

- EIRs are an important element of the project BIM implementation strategy because they are used to set out clearly to the bidder what models are required and what the purposes of the models will be.
- EIRs will be written into the BIM protocol and implemented through the BEP.
- EIRs are key documents with regards to communicating information requirements as well as establishing information management requirements.
- EIRs will act as a good basis from which to review the contents of the tenderer's BEP, confirming its completeness.

The core content of an EIR document should be divided into the three categories of information shown in figure 4.1.

The EIR document sets out some high-level guidance for client organisations to use when planning to procure an asset. There are potentially several key pieces

Technical	Management	Commercial
❍ Software platforms ❍ Data exchange format ❍ Coordinates ❍ Level of detail ❍ Training	❍ Standards ❍ Roles and responsibilities ❍ Planning the work and data segregation ❍ Security ❍ Coordination and clash detection process ❍ Collaboration process ❍ Health and safety and construction design management ❍ Systems performance ❍ Compliance plan ❍ Delivery strategy for asset information	❍ Data drops and project deliverables ❍ Client's strategic purpose ❍ Defined BIM/Project deliverables ❍ BIM-specific competence assessment

Figure 4.1: Information categories for EIRs (CIC, 2013a)

of information required that will be specific to the organisations and which the CIC (2013a) document does not cover. It should also be pointed out that the document has been developed for specifically vertical assets (building) and does not include guidance on horizontal assets such as bridges, tunnels and other infrastructure.

As mentioned previously, apart from providing guidance on what sort of information the employer should ask for by classifying them into the three categories shown in figure 4.1, it also suggests that the methods to be used for management of this information should be spelt out at the initial stages. In addition, the capabilities of the main contractor and the supply chain should also be sought at the tendering stage itself.

So, essentially, the EIR (as developed by a procuring employer organisation) should include the following:

- content that will be of use to the employer's organisation during and after the asset design and build phase
- the format of the contents sought when it is delivered to them
- even before delivering the specified contents, their generation, storage and management through the different stages of a project
- specification of the information delivery (i.e. data drop) points along the project stages

In addition, the *EIR guidance document* (CIC, 2013a) also includes suggestions on the following:

- procurement processes
- PQQs (Pre-qualification Questionnaires)
- details of what sorts of issues employers should consider when developing their EIR specification documents

The EIR guidance document is based on experiences from the early adopter projects that the BIM Task Group were involved in. The BIM Task Group's recommendation is that the tender questionnaire (or PQQ) should separate out the BIM information processing from the asset design and specification for evaluation. These questions should never overlap.

For detailed guidance on the three categories of core content and guidance mentioned in figure 4.1, one should refer to the EIR document (CIC, 2013a). The four points mentioned above are what EIRs should be about, and it is quite possible to cover these points with a different set of organisation-specific guidance that takes the lead from the CIC document but does not entirely follow it verbatim.

Another way to look at the development of EIRs is essentially to keep the end in sight, right from the beginning. So, the more one is under control of what the ultimate goal of the project may be as early as possible in the project life cycle, the more likely it is that those goals will be achieved. Therefore, a slightly modified version of the acronym EIR could well be *End in Review.*

4.3 PAS 1192: Part 2

4.3.1 Introduction

As mentioned earlier, this and the following three chapters deal with Task Group publications and are intended to provide guidance on their key aspects from a practitioner's perspectives. It is not the intention here to give a detailed commentary on every single section of these documents but to identify those parts that are most relevant to the common practitioner and provide guidance on how best to interpret and use the documents in the implementation of Level 2 BIM in project life cycles.

This section will cover the most comprehensive and arguably the most important of the documents published so far: PAS 1192:2:2013. This document was published in February 2013 and its successor PAS 1192:3:2014 has also now been published. Part 2 deals with essentially the asset delivery life cycle from the definition of a 'need' to building an asset to its handover phase. It lays down details on how information is shared in a collaborative environment called CDE, mentioned in chapter 3, as well as specifying in very specific terms the key *data drop* points when the different types of information should be shared. The whole information delivery life cycle is broken down into five key stages and detailed guidance provided on *who* drops *what* information at *what stages* and how the life cycle progresses to subsequent stages by addressing PLQs at these key stages. As stated above, there are other details included in the document that will be omitted for brevity; these are obviously available by consulting the document.

4.3.2 Structure

PAS 1192: Part 2 is organised into ten chapters, out of which six (chapters 5–10) are the main ones structured around the five key stages of the information delivery life cycle. These five stages are:

- assessment and need
- procurement
- post-contract award
- mobilisation
- production

There are several other chapters in the document that deal with the background material before delving into the key six chapters (5–10). It is recommended that one considers and understands figures 3 and 4 of the document (reproduced here in figures 4.2 and 4.3 for easy referencing) before plunging into details of the stages in the latter part of the document. These two figures give a snapshot of various other documents that PAS refers to, and it is essential to have a broad awareness (if not a full understanding) of the scope of these other documents and the relationships between them. One of the challenges of using the documents produced by the CIC (Task Group) so far is that these are a number of rather discrete, disjointed publications and the lay reader could find it quite difficult to link them together. The flow chart of figure 3.8 has been developed with this particular issue in mind. Hopefully, having an understanding of this flow chart along with figures 4.2 and 4.3 helps to obtain a clear and holistic view of the processes enshrined in these documents.

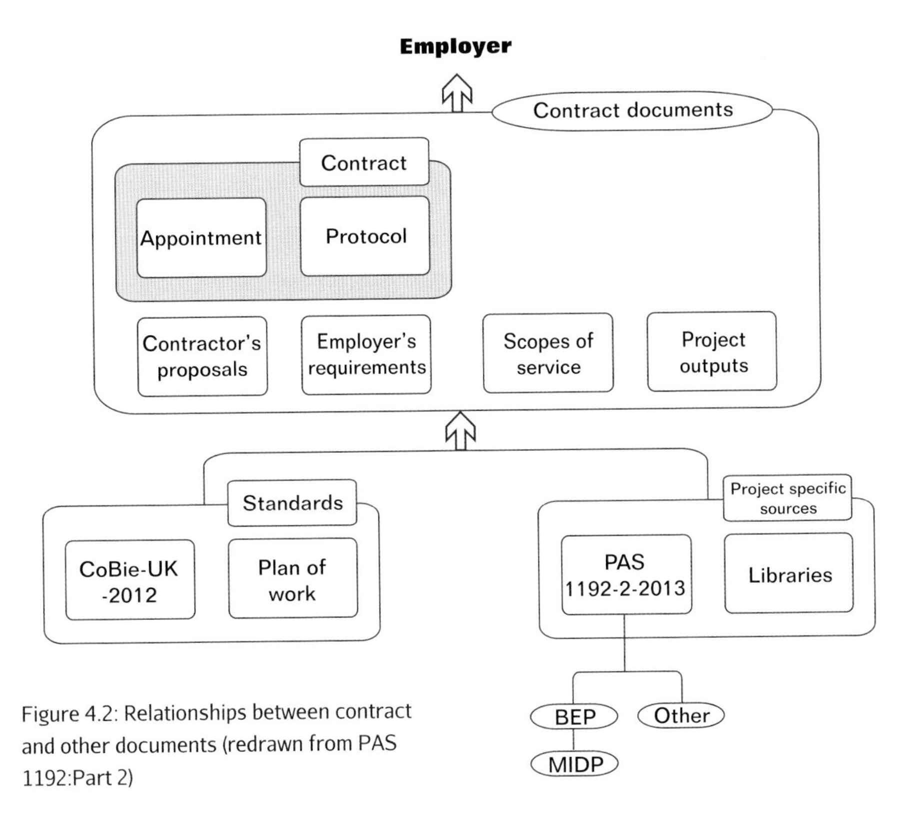

Figure 4.2: Relationships between contract and other documents (redrawn from PAS 1192:Part 2)

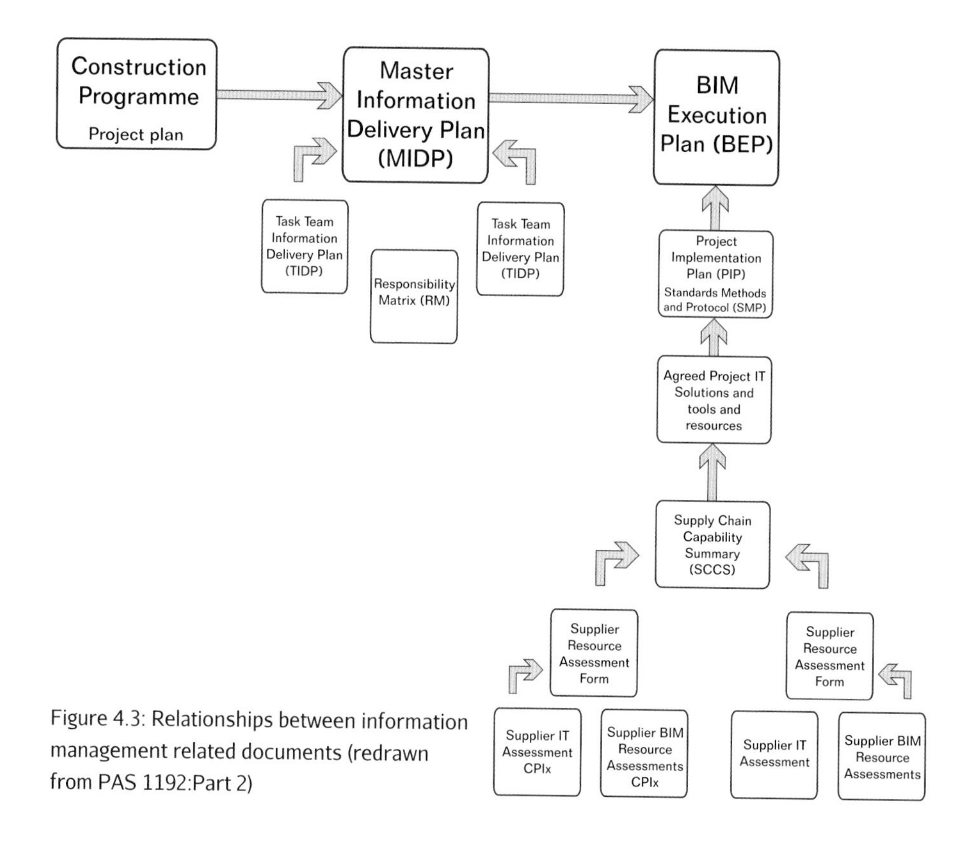

Figure 4.3: Relationships between information management related documents (redrawn from PAS 1192:Part 2)

It is worthwhile spending some time eliciting the message enshrined in figures 4.2 and 4.3. The first point to note in figure 4.2 is that it contains four of the 'shrink-wrapped' documents published by the Task Group specifically mentioned in figure 3. These are the EIRs, BIM protocol, Scope of Services within the *Contract documents* box and PAS 1192:Part 2, which appears in the *Project specific sources* box. There are clearly several additional documents also mentioned that are relevant and which will be required to be used for any BIM-enabled project.

Meanwhile figure 4 of PAS 1192:Part 2 (figure 4.3) illustrates the relationship between various documents mentioned in figure 3 of PAS 1192:Part 2 (figure 4.2).

If one looks closely at figure 4.3, it will be evident that the relationships actually mirror the workflow (with associated documents) mentioned earlier in figure 3.8. Starting at the bottom right end of figure 4.3, the documents mentioned here are outputs of the supply chain's IT capability assessments as stipulated in the EIR document. Moving upwards, one reaches the supply chain capability summary document by aggregating the capability documents of all members of the supply chain. Moving further up, the 'tree' takes two routes, one of which lists the responsibilities of all stakeholders in a responsibility matrix and the other one progresses towards collecting information on various aspects of project IT solutions, tools and resources as well as the PIP (Project Implementation Plan), which eventually forms part of the

BEP document. The detailed methods of design management and the specifics of the procurement strategy and documentation will need to be referenced in detail for actual delivery. These will be described in the PIP and contract documents (PAS 1192:Part 2, p. 10). Finally, at the top of the tree sits the MIDP, which is an aggregation of all the TIDPs (Task Information Delivery Plans) linked with the project plan. The completed MIDP finally then feeds into the BEP document, thereby concluding the collation of data for information management for the project.

Table 1 – Information modelling maturity Level 2

Enabling tools	Many software solutions in combination with many variable interoperable capabilities. Design through manufacture and construction. Discipline-based production/analysis software. File-based collaboration and library management.
BSI Standards	Available: • BS 1192:2007 • BS 7000-4:1996[A)] • BS 8541-1:2012 • BS 8541-2:2011 • BS 8541-3:2012 • BS 8541-4:2012 • PAS 1192-2:2013 • PAS 91:2012 To be developed: • PAS 1192-3 • BS 1192-4
CPI/BSI documents	Available: • A standard framework and guide to BS 1192:2007 Under development: • CPIx Protocol • CPI Uniclass (unified) To be developed: • CPI Uniclass supporting guidance
Other documents	Under development CIC Scope of Services for the Role of Information Management, First Edition, 2013 To be developed: • Early adopters learning report • Institutional plans of work • CIC BIM Protocol, First Edition, 2013 • Employers Information Requirements • Government Soft Landings (policy title to be confirmed)

NOTE 1 This table has been developed from the diagram shown in the Building Information Modelling (BIM) Working Group Strategy Paper, published in March 2011.

NOTE 2 All the above documents will be available from BIM Task Force website at http://www.bimtaskgroup.org.

A) Revision in preparation.

Figure 4.4: Standards, tools and other documents for information management for Level 2 BIM (reproduced from PAS 1192:Part 2)

There is just one more source of information in this document that deserves attention at this stage. This is table 1 of PAS 1192:Part 2 (reproduced in figure 4.4 here for convenience), which gives the reader an idea of the plethora of standards, enabling tools and other documents one needs to be familiar with and potentially use in relation to implementing Level 2 BIM in a project.

It should be pointed out that BS 1192:2007, in particular, should be studied carefully and the key concepts understood before moving on to PAS 1192:Part 2. In particular, the concept of CDE as well as the four phases of each document (WIP (Work in Progress), shared, published and archived) is a legacy from BS 1192. There is an auxiliary document, Guide to BS 1192 (BSI, 2010), that helps to understand some of these key concepts. As shown in figure 4.5, a set of Plain Language Questions (PLQs) are applied at each stage as a test to progress to the next stage. These PLQs will depend on and be developed from the EIRs as specified right at the start of the project. Examples of these PLQs for different stages can be found within the 'Task Group Labs' section of the BIM Task Group website (www.bimtaskgroup.org).

4.3.3 Summary

To summarise, here are some key points to note about PAS 1192: Part 2:

- It is the successor to BS 1192:2007, and the requirements within this PAS build on the existing code of practice for the collaborative production of architectural, engineering and construction information, defined within BS 1192:2007.
- It provides the specifications for information management for the capital/delivery phase of construction projects using BIM.
- It is the predecessor to PAS 1192: Part 3 – the specification for information management for the O&M phase of a built facility.
- PAS 1192: Part 2 focuses specifically on project delivery, where the majority of graphical data, non-graphical data and documents, known collectively as the PIM, are accumulated from design and construction activities.
- Commencing at the point of assessment (for existing assets) or statement of need (for new assets) and progressively working through the various stages of the information delivery cycle, the requirements within this PAS culminate with the delivery of the as-constructed AIM. This is handed over to the employer by the supplier once the PIM has been verified against what has been constructed.

Figure 4.6 illustrates the flow of data between the *push side* of the supply chain and the *pull side* of the client/employer. The *push side* sits above the thick black line, which is essentially the CDE. This is an important diagram to be understood before getting into the details of different components of the information delivery life cycle stipulated by this PAS. The rest of the PAS deals with the transactions above the line, i.e. the mechanisms and transactions within the CDE.

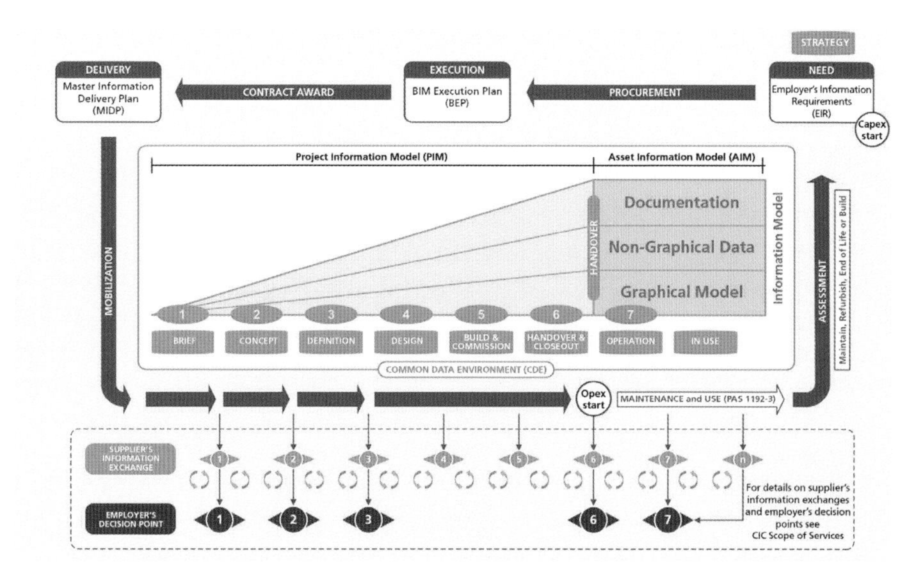

Figure 4.5: Information delivery life cycle (reproduced from PAS 1192:Part 2).

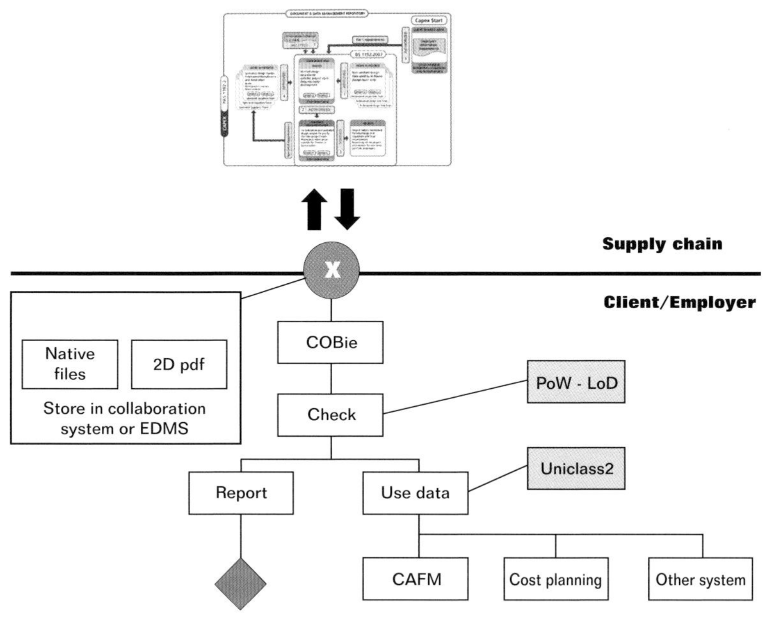

Figure 4.6: Data transactions between the supply chain and the client/employer (redrawn from BIM Task Group website, www.bimtaskgroup.org)

The purpose of figure 4.6 is to illustrate the core drivers of data transactions throughout the key stages of the project from inception to handover. The bottom right-hand part of the diagram essentially shows the end use of data produced up to the handover stage by indicating that the data delivered at handover should be in a format (e.g. Uniclass 2) that could be seamlessly input to the CAFM as well as other systems used in the O&M of an asset. Moving up, the diagram shifts up the abstraction hierarchy of the data produced by the supply chain. For example, the top left (below the thick black line) part of the diagram shows the different components of the data produced at handover in terms of native files of models, 2D PDF cuts of the 3D models and other non-graphical documents. Alongside this set of data, one can see the COBie output sheets that would also be produced, which will be put through checks for levels of details (linked to the stages of appropriate PoWs/CIC work stages) as well as consistency and accuracy. They are then passed down the chain for use in the O&M stage, in addition to generation of various other kinds of report, as required, at different information exchange/data drop points. This drives the whole process by addressing different PLQs by the employer at these stages. The progression from one stage to the next would depend on the answers to these PLQs.

As mentioned earlier, BS 1192:2007 laid down the foundations of data management in a collaborative environment. Figure 4.7 (based upon BS 1192:2007) shows the idea of any project document moving through the stages of *WIP, shared, published documentation* and *archive*. The same approach has been adopted in PAS 1192:Part 2.

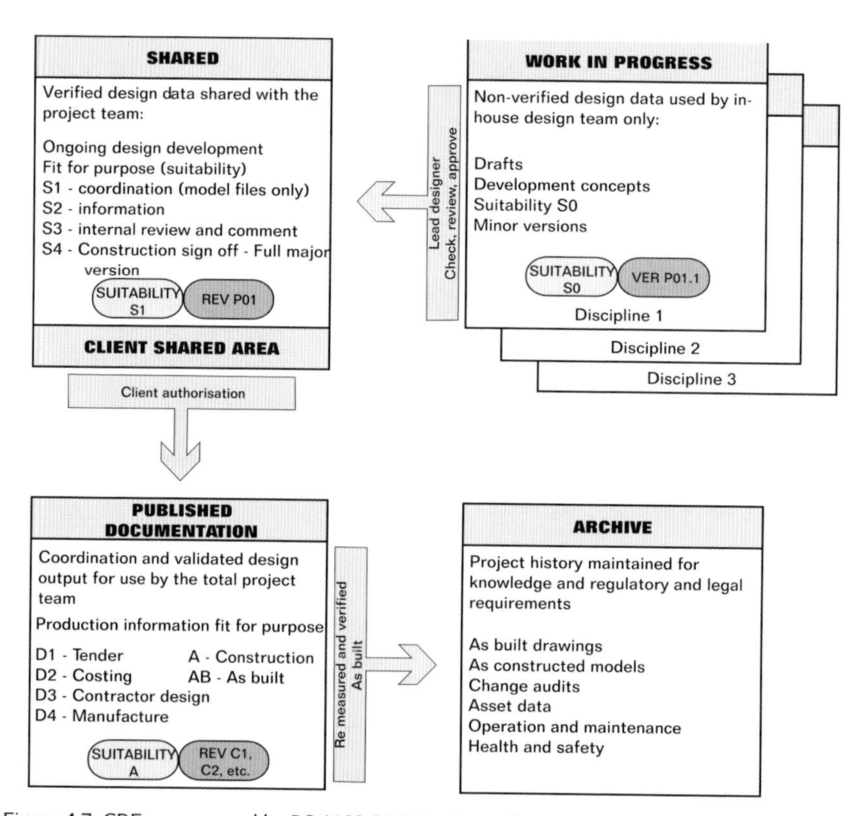

Figure 4.7: CDE as proposed by BS 1192:2007 (redrawn from BS 1192:2007)

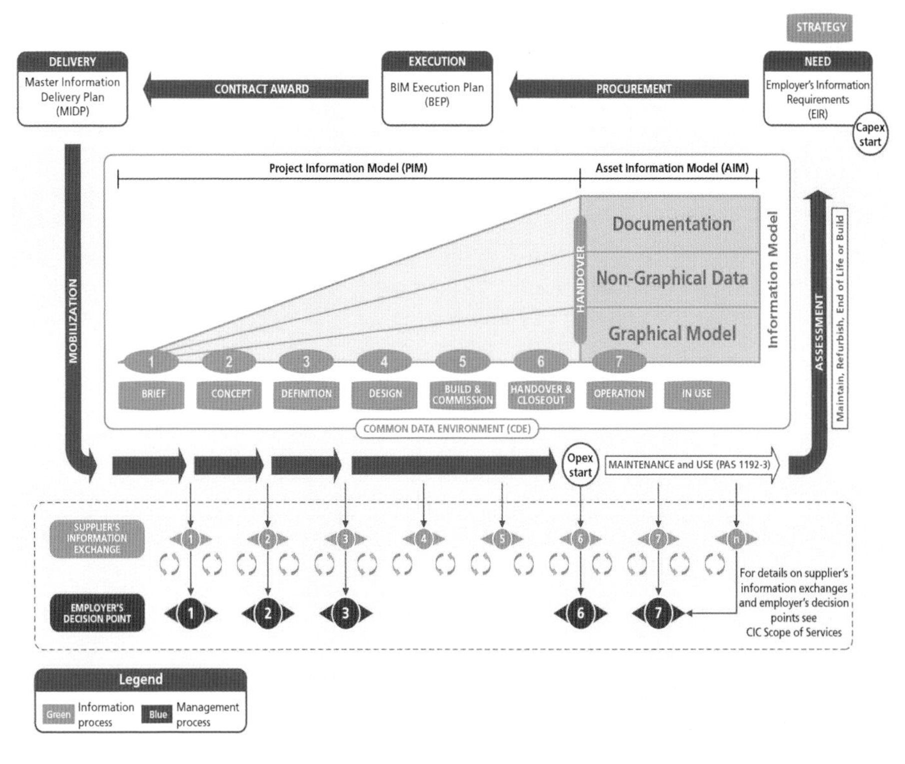

Figure 4.8: Information delivery life cycle (reproduced from PAS 1192:Part 2)

The information delivery cycle (repeated here as figure 4.8) and the project stages described in this PAS begin at 'Capex start' (top right corner of figure 4.8) and end at handover or Opex (Operations and Maintenance) Start. In order to reach the handover milestone, the life cycle progresses through the five stages of Assessment of Need, Procurement, Contract Award, Mobilization and Production culminating in Asset Information Model (AIM) as shown above.. Handover may also coincide with the 'Opex start' milestone. The information management issues at this point onwards (i.e. beyond Handover) are not within the remit of this PAS, but are within that of its successor, PAS 1192:3:2014.

4.4 Data drop points

As mentioned above, the information management life cycle proceeds through a number of data drop points. These data drop points are indicated on the life-cycle diagram shown in figure 4.8 as green circles with diamonds for suppliers and shown in red for the employer. This section provides a brief overview of the key data drop points. The overview provided here is simply to provide a basic understanding of the key drop points without going into the details of how the cycle progress from one drop point to the next. This will be dealt with in later sections.

4.4.1 Data drop 1

This is the first key information exchange point in the life cycle and at this point, the model essentially represents REQUIREMENTS and CONSTRAINTS.

4.4.2 Data drop 2

At this key information exchange point in the life cycle, the model essentially represents an outline solution. However, this drop point consists of two *subpoints*, 2a and 2b. The client's outline solution is labelled as data drop point 2a, whereas the contractor's response is 2b. The idea here is to be able flag the difference between 2a and 2b, which will determine whether alternative solutions should be explored or perhaps an effort may be made to minimise the difference depending on its magnitude.

4.4.3 Data drop 3

This is the drop point that signifies the end of the design phase and when the model represents construction information.

4.4.4 Data drop 4

This drop point essentially signifies the end of the construction phase and the model should contain all the O&M information ready to be handed over to the employer. This is clearly a key stage marking the end of the involvement of the contractors and other members of the design and construction supply chain. It is, therefore, important that the information contained in the model is as complete as possible and in a format (COBie) that can be seamlessly fed into the employer's CAFM and other systems.

4.4.5 Data drop 5 (and subsequent drops)

This drop point deals with post-occupancy validation information. This is not dealt with in PAS 1192:Part 2 and is part of its successor document.

4.5 Stages of the information delivery cycle

The following sections describe the key five stages of the information delivery life cycle from the inception of the project to handover, as covered by PAS 1192:Part 2.

4.5.1 Assessment and need

This stage is essentially centred on EIRs. Once the employer's team has established a need for the asset, the details of requirements for different kinds of information are specified at this stage. The EIR guidance document gives details of different kinds of information that should be included in an EIR. However, this section of the PAS 1192:Part 2 also gives some higher-level guidance on EIRs. Arguably, there is some repetition of material between the documents, but in defence, it can said that the PAS

1192:Part 2 brings together, in a holistic manner, the entire information procurement and delivery processes, whereas the separate documents (e.g. EIRs) give more in-depth details of the specific areas that they cover. For example, section 5.1.2 of PAS 1192:Part 2 states that definition of the information exchange and collaborative working requirements shall be undertaken in parallel with other procurement and project definition activities. Information exchange and collaborative working requirements are described in the EIRs, which form part of the employer's requirements and will in turn be incorporated by a supplier into a PEP. The contents of the EIRs are aligned to employer decision points that in turn will coincide with project stages.

Section 5.1.2 of PAS 1192:Part 2 goes on to say that 'The EIRs shall be consistent with other appointment and contract documents in use on the project, which in turn should be aligned with industry standards such as the RIBA Plan of Work or APM Project Stages. Information requirements set out in the EIRs shall only provide enough information to answer the "Plain Language Questions" (PLQs) required at a particular stage, at an appropriate level of detail'. This is not new material and is available in the EIR guidance document, but it has been cast here in terms of the overall information delivery life cycle in PAS 1192:Part 2.

4.5.2 Procurement

In this section, similar to the previous stage of assessment and need, PAS 1191:Part 2 puts things into perspective for the overall information delivery cycle in relation to the procurement stage of the project.

The main objective of this stage is give the employer an opportunity to assess the feasibility of the requirements laid down in the EIR as well as to establish if there may be a need to adapt them, if required, based on the capabilities of the supply chain.

In order to achieve this, the pre-contract BEP is first produced by the bidders in conjunction with the EIR. The bidders submit this document on behalf of their whole supply chain. Alongside this, the PIP also needs to be submitted. Supplier BIM and IT assessments should be carried out and submitted together with a summary of the supply chain's capability vis-à-vis IT and BIM, as stipulated in the EIR.

In summary, this stage spells out all the various pieces of information that need to be in place on behalf of the entire supply chain for the employer to assess and select the preferred supplier who will then negotiate further with the employer to agree the final set of documents that will form the basis for putting together the post-contract BEP later.

4.5.3 Post-contract award

By this stage, the agreed EIRs as well as the pre-contract BEP are in place. So, the ground is set for the production of the post-contract BEP. This is the stage when all the ingredients of the MIDP are in place and can now be produced. Finally, the roles for the project delivery team and their responsibilities as well as authorities need to be agreed, documented and signed off.

4.5.4 Mobilisation

This is the stage when the project delivery is supposed to *test* the solutions that form part of the information delivery infrastructure before any active project activities (e.g. design) are kicked off. This includes ensuring that all the required pieces of software, hardware, preparation and agreements on contractual documents and the key information management processes are in place. In addition, the team also needs to ensure that all the required skillsets for implementing the processes are in place. Finally, the signed-off BEP needs to be communicated to all stakeholders in the project delivery team. It is all too common to find that the project delivery team *gets down to business* without a proper mobilisation phase and consequently discovers that parts of the infrastructure do not actually work as planned, thereby causing delays and escalated costs. PAS 1192:Part 2 makes this stage an explicit part of the whole process and recommends that the actual project execution not be implemented without thorough testing of all the required infrastructure and documents being in place.

It should be pointed out, however, that although this document does not explicitly address the issue of interoperability between software to be used on the project, it does provide guidelines on what formats data should be made available. It also suggests that the team should clearly understand the limitations of different software to interoperate before their actual use gets under way and a strategy for interoperability is agreed before the software is in place.

Finally, this section suggests that the training needs of the team should be assessed and that a plan for implementing them should be agreed at this stage. The training and education issues are covered in detail in chapter 7 of this book.

4.5.5 Production and role of CDE

This is, by far, the longest section of the document and deals with some of the most important concepts such as CDE. As mentioned previously, CDE is based on guidance laid down in this PAS's predecessor, BS 1192:2007. However, this section details the extensions to those ideas. In a nutshell, the CDE is a managed PE (Project Extranet) environment with added checks for quality as well as infrastructure in place for information exchange in much more integrated and seamless ways in contrast to the traditional PEs, which are essentially a repository for different kinds of project documents. Figure 4.9, reproduced from PAS 1192:Part 2 for convenience, clearly shows the part that comes from BS 1192:2007 (as evidenced from comparing it to figure 4.7) alongside all the additional components that have been added to it to facilitate BIM-enabled information sharing and exchange. As mentioned earlier, BS 1192:2007 was basically developed for CAD-based data exchange.

This is the stage where development of PIM starts in earnest based on the MIDP. PIM in Level 2 includes a set of all the building information models along with non-graphical information and other relevant documents that may add additional details of elements in the models. As explained previously, PIM evolves gradually with the different stages, as shown in figure 4.9. As the process moves from one stage to the next, the volume of

Architect's issue to be SHARED

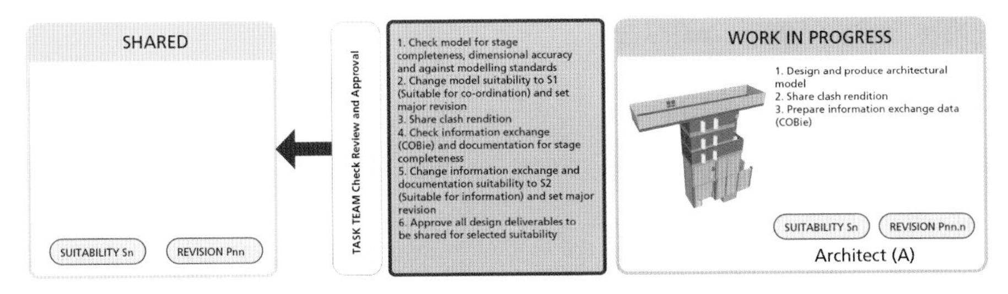

Structural Engineer's issue to be SHARED

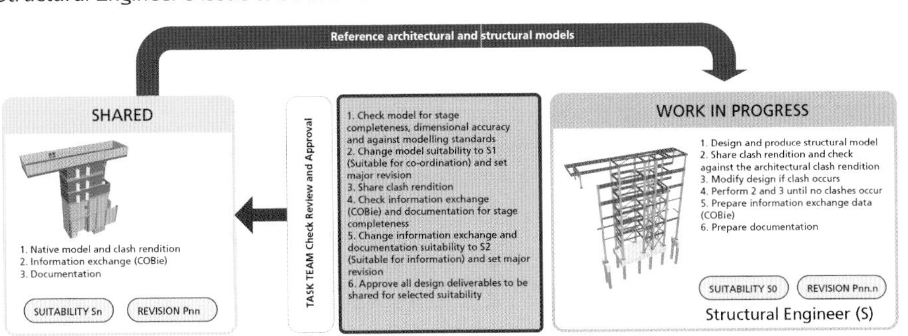

MEP Engineer's issue to be SHARED

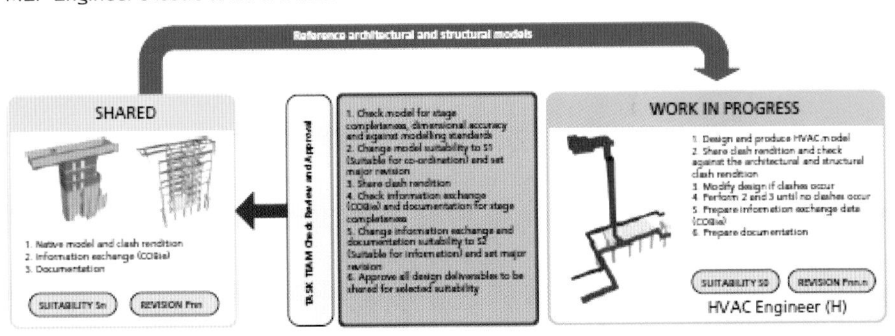

Figure 4.9: Movement of data between different stakeholders through the CDE (reproduced from PAS 1192:Part 2)

information keeps on increasing, finally ending in the fully developed PIM. This progress from one stage to the next will depend on the information exchanges and the data drop points explained earlier, which will, in turn, inform the employer to take decisions to progress the process based on answers to the PLQs at each of these decision points. In parallel, the CDE's role is to ensure that the data generated is accurate, consistent and clean. This includes carrying out tests to ensure that the data complies with certain laid down constraints. This is where CDE plays a major role. As shown in figure 4.9, every piece of data has to go through gatekeepers and gateways before it can be shared with the rest of the project stakeholders. CDE then puts the data through further checks before authorising publication and finally after even further checks, moves the data to be archived. As one can see in figure 4.10, there are six gateways altogether specified by the PAS 1192:Part 2. In summary, this phase of the information delivery cycle (i.e. production) is largely driven by the management functionalities of the CDE. Therefore, a clear understanding of the roles that the CDE plays and the benefits it accrues to the whole process is essential. In practical terms, here are the key points one should be mindful of when setting up the CDE:

- It provides more than just a central repository of data and therefore using a classical PE environment for a CDE is not appropriate.
- It ensures that the data generated at different stages complies with the requirements of the employer's needs by carrying out checks at different stages for several key criteria such as completeness, suitability of the model itself and COBie 2012 UK (BS1192-Part 4, 2014) dataset compliance, among others.
- It maintains the models in a federated structure, thereby maintaining the ownership of models with the originators whilst allowing others to use them as reference material for their own activities. This is the essence of Level 2.
- It separates the unapproved data for the specialist subcontractors within the WIP section. There may be a change of ownership for the data introduced by these specialist subcontractors. It is critical to ensure that the subcontractors do not alter the original design elements but introduce their own models, which should be clearly demarcated by the CDE to avoid conflicts and confusion.
- In the archive section, the CDE records the completions of all activities, thereby maintaining the full progress information along with full details of any change orders to facilitate any audit trails for conflicts that may arise at a later date. In addition, the archive section also stores the final as-constructed information from the published section after proper checks.

It is therefore clear that the CDE plays a central role in ensuring that the information delivery cycle progresses from one stage to the next in as seamless a manner as possible. Therefore, the design and installation of a robust and fit for purpose CDE is paramount to the success of the information delivery cycle as stipulated by PAS 1192:Part 2. As a practical guide, one can take CDEs to be *managed PEs*. As far as the author knows, there are no commercial CDE products being sold on the market at the

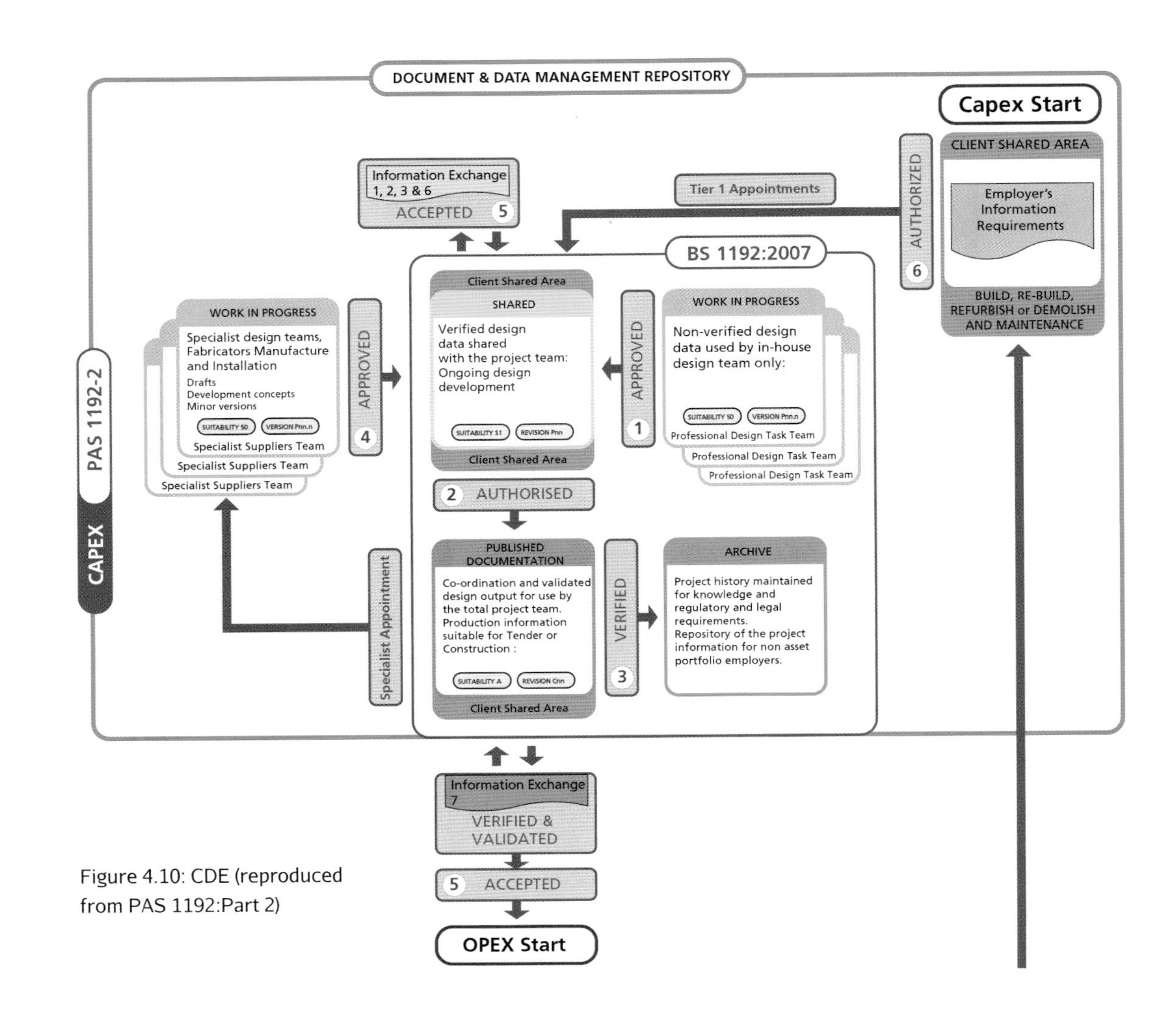

Figure 4.10: CDE (reproduced from PAS 1192:Part 2)

moment. But, the practical solution at this point seems to be a *customised PE* that can accomplish the key functionalities mentioned earlier.

PAS 1192:Part 2 deals with all these issues in some detail and although sufficient detail has been provided here for practical purposes, someone keen on full details is directed to the actual document.

4.5.6 AIM model – handover of information for Opex

Status codes, file and layer naming conventions and classification

This section provides an overview of some of the other key issues covered by PAS 1192:Part 2 and which are of direct relevance from a practical perspective. The first one is the concept of status codes. As has already been stressed, one of the key drivers for effective and seamless information exchange is common data formats and standards used by all parties. It is in this regard that this PAS specifies the guidelines on status codes as well as file and layer naming conventions that all parties should use. Even if only one party does not adhere to these standards, the whole process will fall down. This obviously includes the subcontractors and the entire supply chain. Therefore, whenever a piece of data is generated or passed on, whether it is in the WIP or the shared or published areas of the CDE, they should all be assigned a universally agreed status code. There are 16 codes specified by the PAS that should be used by all the parties for every piece of data that is stored in the CDE. Similarly, for the same reasons, a standard naming convention for all files and layers used in the models must be used by all parties concerned. This, of course, should not come as a surprise to anyone with any CAD or any software development experience. Standard file naming conventions is one of the basic concepts in programming methods. This PAS provides comprehensive guidance on naming conventions for files and layers. Finally, classification plays a paramount role in information exchange for the same reasons as mentioned for status codes and naming conventions. Note 2 of section 9.10 of PAS 1192:Part 2 states that 'A classification system provides a common terminology and structure to which all project documents and information can be related. The use of classification is required in information exchange and in the COBie-UK-2012 templates.' Therefore, the hugely important roles that these seemingly simple (and often ignored) issues perform in the overall information management and exchange process should be absolutely clear. PAS 1192:Part 2 specifies five different classification systems to be used for different kinds of information. These are NRM 1, 2 and 3, CESSM and Uniclass. It is vitally important that sufficient familiarity and experience is gained in these classification systems for Level 2 compliance.

COBie

Much has been made of the importance of information exchange standards in the overall information exchange process. This section deals with two of these standards (actually one is a subset of the other) that the construction industry uses, and it is therefore important to appreciate their significance in the information exchange processes. The standard that the BIM Task Group has specified for the final handover information is called COBie. COBie is the specification for facility management within the IFC data model. COBie is, therefore, a subset of IFC. A comprehensive discussion on IFC or COBie is out of the scope of this book. However, some introductory material is provided here to put things into perspective.

Here are some key points about COBie:

- COBie stands for Construction Operations Building Information Exchange
- COBie is a data schema for holding and transmitting information around handover to support the client's ownership and operation of a facility
- Technical specification for COBie is the IFC facility management handover model view definition

As part of Level 2 BIM compliance, the UK Government has stipulated that from 2016, the main deliverables for all public sector projects will be:

- COBie UK 2012 dataset
- Level 2 BIM models
- 2D PDFs of the drawings

COBie is formally defined as a subset of IFC, including initially the FM-10 handover model view definition and latterly the full COBie. It can also be represented in spreadsheets or relational databases. It was developed by the US Army Engineer Research and Development Center with the National Aeronautics and Space Administration and the Veterans Association as a response to the perception that specifications for O&M and other documentation did not allow these owner organisations to take on effective management immediately on handover. The COBie data schema has subsequently been generalised for international use by ensuring that classification and unit systems can be specified.

During traditional projects, most of the information required by COBie is already delivered in unstructured form. COBie gives the opportunity to input critical data just once, allowing it to be reused in many outputs, be tested in many ways and be delivered to many applications including facility management and asset management systems.

The COBie process requires that there are progressively more complete contractually required data deliverables throughout the design and procurement stages, culminating at the handover point. Any deliverables after the point of handover are expected to consist only of corrections and data obtained during post-occupancy assessments.

COBie may be created using one of the following three methods:

- use of COBie-compliant software
- development of custom transformations of existing data into a COBie-compliant file
- direct use of the COBie spreadsheet format

At the moment there are strong arguments for and against using COBie as a data format for Level 2 BIM. Some of the criticism is based on experience with the early adopter projects, which demonstrated that the amount of effort required to create some of these COBie spreadsheets was unacceptably high. However, with the availability of automated tools to generate the COBie sheets, the effort required should be much reduced.

IFC

IFC is the industry-supported set of standards for representing and exchanging information about building objects and entities. IFC is an International Organization for Standardization standard data schema for holding and transmitting facility information throughout the facility life cycle. IFC has been developed over the past several years by buildingSMART, a not-for-profit consortium with chapters in the US, UK, Europe, the Middle East and Africa, the Far East and Australia.

The objective of COBie is not to change the type of information that is required, just to standardise the format of that information to save you, and the buildings' owners and occupants, having to rekey this information multiple times.

IFC was originally envisaged to facilitate interoperation between different software systems. It was acknowledged for several years that the industry suffered huge losses of amounts of time (and consequently money). A survey carried out by NIST in the USA puts the annual loss to the construction industry due to lack of interoperation at well over US$15 billion per year.

Universal interoperation between software can be achieved by one of two ways:

- using all software from one vendor
- using software from different vendors that can exchange data using industry-supported standards

The trouble with using all software from one vendor is that there is no vendor who produces software which can support all activities throughout the life cycle of a building design, construction and O&M.

So, the only other option is to use software available from different vendors that can 'talk' (exchange data between each other using industry-supported standards) to each other. This is where IFC comes in. There is a growing number of software products supporting different aspects of a building life cycle that can 'talk' to each other using IFC as the standard for the data model used by them. However, there is work to be done because IFC suffers from several limitations. So for the moment, the only solutions are to use software from the same vendor as much as possible, or use IFC schemas to exchange information between IFC-compliant software if at all possible.

Limitations of IFC

- Data loss can happen both in importing from and exporting to the IFC format.
- For the IFC model to facilitate full interoperability between applications, it would have to be a superset of all their data models, which would

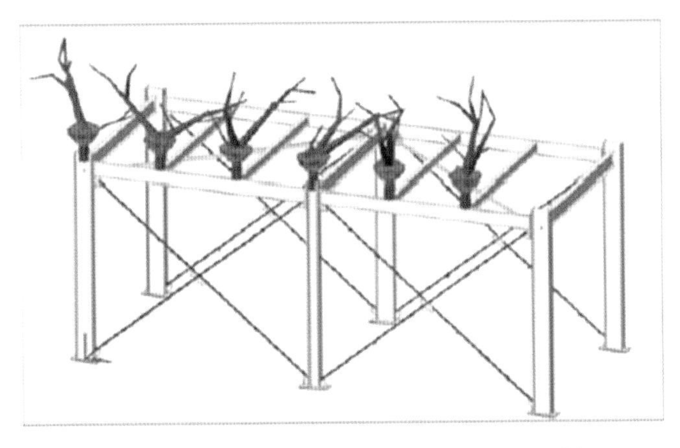

Figure 4.11: Issues with incorrect data mapping using IFC (reproduced from Lipman, 2006)

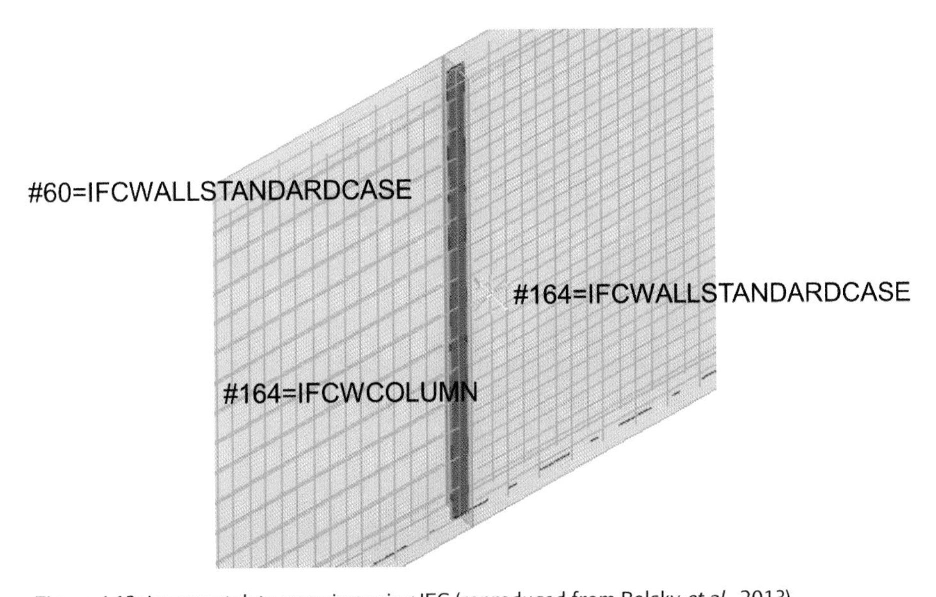

Figure 4.12: Incorrect data mapping using IFC (reproduced from Belsky *et al.*, 2013)

be a near-impossible task. It is important to keep this in mind so that expectations from the IFC do not exceed what is realistically possible.

In the example shown in figure 4.11, an incorrect mapping was done by the translator as there was no schema provided in IFC on connection details and was arbitrarily mapped to a 'plant' object! In the example shown in figure 4.12 (Belsky *et al.*, 2013), there are two wall panels and a sealing strip between them. In the IFC model, the wall panels are represented by the IfcWallStandardCase entities and the sealing strip is represented by the IfcColumn entity. This is a typical IFC export representation, derived from a popular commercial BIM tool. In the exported file, there is no semantic construct to represent the precast joint between the walls.

4.5.7 Levels of model definition

Another key concept addressed in this section of PAS 1192:Part 2 is the Level of Detail in the model at different information exchange (data drop) points. Clearly it does not make any sense to provide more detail than required at the different exchange points. But it is also equally important to define a universally agreed set of standards for the Level of Detail that is followed by all stakeholders. Section 9.8 gives some detail guidance on this mapped onto the different CIC work stages from 'Brief' to 'In Use'. Figure 20 of the PAS 1192:Part 2 gives fairly comprehensive guidance on this.

Section 3.9 has already provided an alternative way of standardising the LOD provided by AIA. Arguably, the AIA LOD definitions are different from those provided by PAS 1192:Part 2. Both approaches have their own benefits. However, arguably the AIA approach is more structured in that it provides a very definitive set of guidance by defining and labelling different LODs in a more precise way. For example, the AIA approach uses a numerical system (100, 200, 300, 400 and 500) for every element for an asset broken down by the different parts of the asset (such as substructure, interior, equipment, furnishings) to define how much detail should be provided for these numerical levels, which makes it much easier to communicate LODs to other parties. There is no link between project or work stages and Level of Detail in the approach, and the reasoning provided is that at any stage, different elements in a model may well be at different Levels of Detail.

Here are the AIA LOD definitions:

LOD 100 The model element may be graphically represented in the model with a symbol or other generic representation, but does not satisfy the requirements for LOD 200. Information related to the model element (i.e. cost per square foot, tonnage of heating, ventilation and air conditioning, etc.) can be derived from other model elements. This LOD maps on to old RIBA PoW Stage B.

LOD 200 The model element is graphically represented within the model as a generic system, object or assembly with approximate quantities, size, shape, location and orientation. Non-graphic information may also be attached to the model element. This LOD maps on to old RIBA PoW Stage B/C.

LOD 300 The model element is graphically represented within the model as a specific system, object or assembly in terms of quantity, size, shape, location and orientation. Non-graphic information may also be attached to the model element. This LOD maps on to old RIBA PoW Stage D+.

LOD 350 The model element is graphically represented within the model as a specific system, object or assembly in terms of quantity, size, shape, orientation and interfaces with other building systems. Non-graphic information may also be attached to the model element. This LOD maps on to old RIBA PoW Stage D/E.

LOD 400 The model element is graphically represented within the model as a specific system, object or assembly in terms of size, shape, location, quantity and orientation with detailing, fabrication, assembly and installation information. Non-graphic information may also be attached to the model element. This LOD maps on to old RIBA PoW Stage E.

LOD 500 The model element is a field-verified representation in terms of size, shape, location, quantity and orientation. Non-graphic information may also be attached to the model elements. This LOD maps on to old RIBA PoW Stage E/F.

Example – light fixture:

100 cost/s attached to floor slabs

200 light fixture, generic/approximate size/shape/location

300 design-specified 2 × 4 troffer, specific size/shape/location

350 actual model, Lightolier DPA2G12LS232, specific size/shape/location

400 as 350, plus special mounting details, as in a decorative soffit

On the other hand, the PAS 1192:Part 2 approach is linked with the CIC stages. One might argue that the PAS approach is more appropriate in that it links it to the different stages and stipulates a universal view for the whole model at a particular stage. There are clearly advantages and disadvantages in both approaches and, at the end of the day, the ultimate choice will be down to personal preferences; the key will be to use the approach as prescribed in the EIR. Section 9.8.4 of PAS 1192:Part 2 stipulates that the level of model definition shall conform to the:

- EIR
- scope of work set out by the CIC Scope of Services (CIC, 2013c), e.g. related to the project stages
- Uniclass classification tables regarding the relationships of systems, products and elements with the specification and the cost plan

Stage number Model name	1 Brief	2 Concept	3 Definition	4 Design	5 Build and commission	6 Handover and closeout	7 Operation
Systems to be covered	N/A	All	All	All	All	All	All
Graphical illustration (building project)							
Graphical illustration (infrastructure project)							
What the model can be relied upon for	Model information communicating the brief, performance requirements, performance benchmarks and site constraints	Models which communicate the initial response to the brief, aesthetic intent and outline performance requirements. The model can be used for early design development, analysis and co-ordination. Model content is not fixed and may be subject to further design development. The model can be used for co-ordination, sequencing and estimating purposes	A dimensionally correct and co-ordinated model which communicates the response to the brief, aesthetic intent and some performance information that can be used for analysis, design development and early contractor engagement. The model can be used for co-ordination, sequencing and estimating purposes including the agreement of a first stage target price	A dimensionally correct and co-ordinated model that can be used to verify compliance with regulatory requirements. The model can be used as the start point for the incorporation of specialist contractor design models and can include information that can be used for fabrication, co-ordination, sequencing and estimating purposes, including the agreement of a target price/ guaranteed maximum price	An accurate model of the asset before and during construction incorporating co-ordinated specialist sub-contract design models and associated model attributes. The model can be used for sequencing of installation and capture of as-installed information	An accurate record of the asset as a constructed at handover, including all information required for operation and maintenance	An updated record of the asset at a fixed point in time incorporating any major changes made since handover, including performance and condition data and all information required for operation and maintenance The full content will be available in the yet to be published PAS 1192-3
Output	Project brief and procurement strategy	Refined project brief and concept approval	Approval of co-ordinated developed design	Integrated production information	Integrated production information. Complete fabrication and manufacturing details, system and element verification, operation and maintenance information Modify to represent as installed model with all associated data references	As constructed systems, operation and maintenance information Agreed Final Account Building Log Book Information gathered as key elements are completed to feed installation information for the later packages	Agreed final account In use performance compared against Project Brief Project process feedback: risk, procurement, information management, Soft Landings

Figure 4.13: LOD (reproduced from PAS 1192:Part 2)

In general, it is probably more pragmatic to use the AIA LOD approach as, arguably, it is more expressive and more granular in that it deals with the elements in the asset rather than the asset as a whole and is independent of the stage of the project. This, in theory, should give the received information much more visibility of the state of development of the particular elements in question. In addition, it is also flexible enough to cope with different Levels of Detail for different elements at any point in time, which is a more realistic and pragmatic scenario. However, to be more comprehensive, it is recommended that both the approaches should be used in tandem, the elemental LOD being covered by the AIA approach and the LOD at different stages guided by the PAS 1192:Part 2 as shown in figure 4.13.

4.6 PAS 1192:3:2014

In March 2014 the successor to PAS 1192:Part 2 was published. PAS 1192:Part 3 deals with the stage after handover at the end of construction. This book is mainly about the design and construction phases, so Part 3 will not be dealt with here. However, a very high level comparison between the processes that the two documents deal with will be given here. The information delivery cycle that Part 3 proposes for the O&M and other phases beyond handover is significantly different from that proposed by PAS 1192:Part 2. Part 3 proposes that as soon as the process enters the O&M phase, there is no fixed sequence of stages that the process follows. In reality this phase is marked by planned and unplanned events and activities, which is so typical of the O&M phase. As is well known, some maintenance activities might well be planned in advance. However, there are several other unplanned activities that may arise due to various unforeseen circumstances.

4.7 Summary

This chapter has summarised the guidance on key aspects of the information delivery cycle as proposed by PAS 1192:Part 2, as well as the EIRs. There are a few aspects of the PAS and EIR documents that have not been covered in this chapter, as the focus has been to identify mainly the aspects that are key from an implementation perspective. The suggestion here is not that these other aspects should be completely overlooked but the ones covered here should first be fully understood before referring to the others that have been omitted here. The main objective of this chapter has been to summarise the aspects of these documents that should give a practitioner a head start in implementation issues. From this perspective, the main aspects of PAS 1192:Part 2 are the information delivery cycle stages and COBie data drop points, status codes, file and layer naming conventions, classification of information and level of model definition. It is proposed that a good understanding of and expertise in these aspects are absolutely paramount to achieving Level 2 BIM compliance.

5

BIM PEP document development

5.1 Introduction

A PEP (Projection Execution Plan) is a document that most 'well-managed' projects have in place and is treated almost as a bible for the project as a central point of reference for different aspects of the project.

5.2 PEPs

A PEP:

- is one of the project manager's key toolkits just like bar charts, CPM/PERT (critical path method/programme evaluation and review technique) schedules, earned value reports, etc.
- captures and helps monitor the key project controls
- is the key reference source for:
    - key project information such as project organisation and hierarchy, including the lines of communication and management, among others
    - key project processes
    - key deliverables at main points/stages of the project processes
    - roles and responsibilities of all stakeholders in the project

According to the CIOB (Chartered Institute of Building), a PEP provides a 'guide to the project team members in the performance of their duties, identifying their responsibilities and detailing the various activities and procedures (often called the project bible)...' (CIOB, 2002).

The APM (Association for Project Management) suggests that a PEP '...confirms the agreements between the sponsor and other stakeholders and the project manager ... [it] documents how the project will be managed in terms of why, what, how (and how much), who, when and where' (APM, 2014).

The OGC proposes that 'The Project Execution Plan is the key management document governing the project strategy, organisation, control procedures, responsibilities, and, where appropriate, the relationship between the project sponsor and the project manager. It is a formal statement of the user needs, the project brief and the strategy agreed with the project manager for their attainment. The scope of the plan will depend upon the size and nature of the project. It is a live active management document, regularly updated, to be used by all parties both as a means of communication and as a control and performance measurement tool' (OGC, 2014).

A PEP document is where all the key aspects of a project are stored in one place. It not only includes implementation issues, but a good PEP also includes a business case for the project and what the initial objectives were for inception. It is important for project team members to understand these in order to deliver a product that is fit for purpose. Of course, the detailed implementation aspects are important too, but these must be driven by the higher-level objectives of the project. As mentioned earlier, the PEP should then include a clear description of project organisation that clarifies the lines of authority. A typical PEP document will contain the following sections:

- project objectives and business case
- risks and opportunities for the project in achieving the objectives
- roles and responsibilities of all team members
- strategies for monitoring and controlling the project implementation including the quality issues
- strategies for handover and O&M are included in some PEPs, but many do not include this and limit themselves to design and construction phases up to the handover stage

5.3 BIM PEPs

Similar to a typical PEP, every BIM-enabled project should have a BIM PEP document agreed and signed off by all stakeholders, right at the start of the project. Contractually, this document becomes an addendum to the contract documents (BIM Protocol, 2013). Therefore, every stakeholder of a project is contractually bound to comply with this document. There are several reasons why a document such as a BIM PEP is essential to ensure that all stakeholders in a project deliver what is expected of them. Introducing BIM in a project usually means bringing in new processes, particularly in terms of information management. In order to successfully manage information in a project, everyone involved in the project needs to sign up to processes and standards in advance of execution of the project. This is a point that has been comprehensively made in earlier chapters. This can only be achieved by careful advanced planning and documenting all processes mapped on to the responsible parties alongside the different stages of the project. Therefore, whenever there is a lack of clarity, dispute or confusion about any aspect of delivery of information throughout the life cycle of the project, the BIM PEP is the document that the project team should rely on for resolution. It is, therefore, not hard to imagine the crucial and important role that this document can play in successful project delivery. It should be pointed out that in the UK, the term BEP is more commonly used than BIM PEP.

5.3.1 BIM PEP development

There are several methodologies proposed for developing a BIM PEP document. One of these, developed at PSU (Pennsylvania State University) in the USA, provides a structured procedure for creating and implementing a BIM PEP as mentioned below. The four steps within the procedure include (PSU BIM PEP, 2010):

- identification of high-value BIM uses during project planning, design, construction and operational phases
- design of the BIM execution process by creating process maps
- definition of the BIM deliverables in the form of information exchanges
- development of the infrastructure in the form of contracts, communication procedures, technology and quality control to support implementation

PSU BIM PEP (2010) suggests that by developing a BIM plan, the project team members can achieve the following values:

- all stakeholders will be able to understand and communicate the strategic objectives behind implementing the BIM on the project
- all tier 1 organisations and their supply chain will understand their roles and responsibilities in BIM implementation
- be able to design an execution process that is appropriate for their business practices and typical workflows
- the plan will help to identify resources, training or other competencies necessary to successfully implement BIM in the project
- the BIM PEP will facilitate engagement with future participants in the project fairly smoothly

As described extensively in earlier chapters, the points mentioned above have echoes of processes outlined in PAS 1192:Part 2. However, the PAS starts off with an assumption that the information delivery cycle will be BIM-driven. The PSU BIM PEP development strategy takes a step backwards and first tries to establish what the main uses of BIM may be in the project at hand before actually developing processes for achieving those benefits using BIM-based processes. This is a subtle difference, but it is important to appreciate it. It is arguably a good idea to first establish what the BIM is going to achieve for the project and how it aligns with business and other goals for the project. The PSU approach also provides a list of uses and goals for the BIM, which is not exhaustive but which is fairly comprehensive. Some of these are given in figure 5.1.

This is probably the most comprehensive description of a procedure for developing a BIM PEP. However, for most practitioners, it will be a case of filling in information in

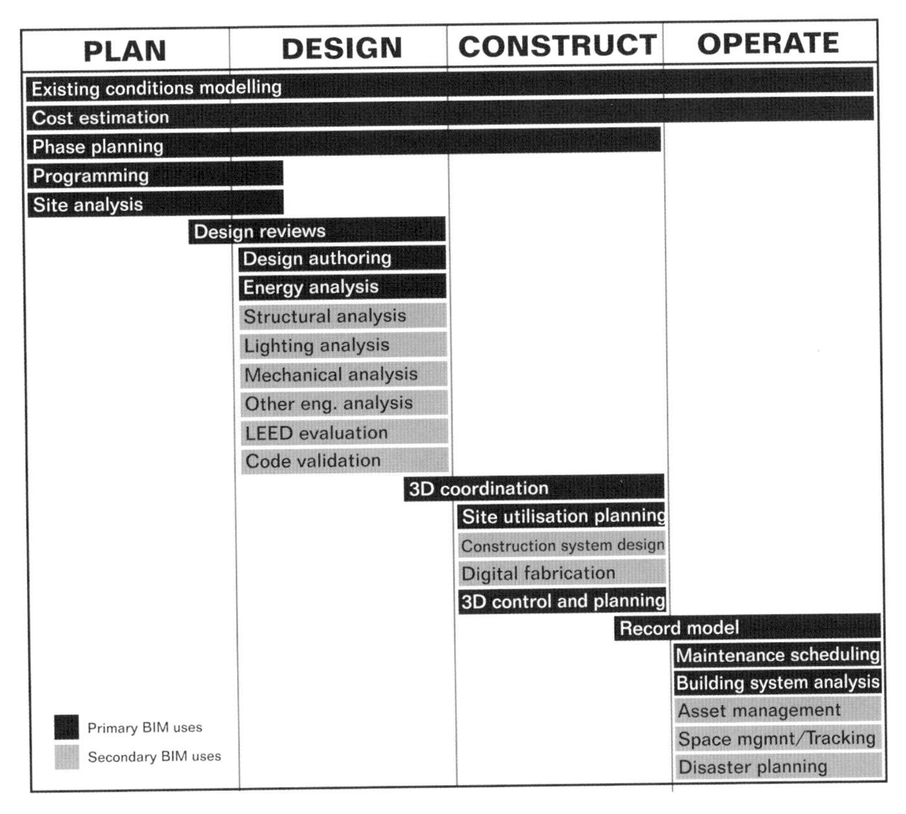

Figure 5.1: Primary and secondary BIM uses (redrawn from PSU BIM PEP (2010) Note LEED = Leadership in Energy & Environmental Design)

pre-designed templates for a BIM PEP. Therefore, for practical purposes, the procedure outlined here is probably irrelevant, but it is included simply to put things into perspective and provide a background as to why the contents of a typical BIM PEP may be required in the first place.

5.3.2 Pre-contract BEPs

PAS 1192:Part 2 introduces the concepts of pre- and post-contract PEPs. This is a new concept not seen elsewhere in the world. In line with this, the CPIC (Construction Project Information Committee) in the UK have recently published their own version of BIM PEP called BEP (CPIC, 2013).

The pre-contract BEP is prepared by the supplier to demonstrate a proposed approach, and the capability, capacity and competence of the firm and its supply chain to meet the EIRs. It will also list all agreed elements as outlined in the Brief, BS 1192:2007, PAS 1192:2:2013, the CPIx protocol and the contract documents. This BEP is structured in accordance with PAS 1192:2:2013.

5.3.3 Post-contract BEPs

CPIC (2013) suggests that the post-contract BEP shall list the agreed targets for responsibility, timely delivery, exchange, reuse and final handover to the clients. It will also list all agreed elements as outlined in the EIRs, the Brief, the BS 1192:2007, the PAS 1192:2:2013, the CPIx protocol and the contract documents. This post-contract BEP is structured in accordance with PAS 1192:2:2013.

There are several other examples of BIM PEPs in use in different parts of the world (PSU BIM PEP, 2010; MIT BIM PEP, 2014). A sample BIM PEP from a building services company is provided in Appendix 3. This sample document shows the depth and breadth of coverage of various issues that are important for successful implementation of any project. In addition, there are several other sample BIM PEP templates available for free downloads from various sources (PSU BIM PEP, 2010). It should, however, be pointed out that these are just sample templates; the exact nature of issues to be covered for a specific project may vary, and organisations should develop their own templates based on these.

5.3.4 Contents of a BIM PEP

Based on the discussion above, one can conclude that a BIM PEP should include everything that is usually found in a 'normal' PEP document, supplemented with additional information relevant to the BIM-based nature of the project. Here is a typical list of items one should expect to find in a BIM PEP document. The first few items are the ones that can be found in any PEP document, whilst the rest are specific to the BIM:

- project objectives and business case
- risks and opportunities for the project in achieving the objectives
- roles and responsibilities of all team members
- strategies for monitoring and controlling project implementation including quality issues
- strategies for handover and O&M are included in some PEPs, but many do not include these and limit themselves to design and construction phases up to the handover stage
- building information model quality control and assurance processes
- building information model LOD
- information exchange standards and data drop points
- BIM and other technology requirements
- model handover standards for facilities management

This is only an indicative list and is meant to serve as a guide and is by no means exhaustive.

5.4 Summary

This chapter has provided a brief background for BIM PEP documents, briefly outlining a procedure for developing them. It was, however, pointed out that instead of following the procedure, most organisations need to develop templates that would address their specific issues, which could be used in different projects. BIM PEPs have a central role in implementing BIM processes and standards in any project and must be used on all BIM-driven projects. Readers have been directed to several sample BIM PEP documents, one of which is also included as an appendix to this book. It was also pointed out that BIM PEPs are generally termed as BEPs in all publications of the BIM Task Group in the UK.

6

BIM protocol, Outline Scope of Services and PII

6.1 Introduction

This chapter provides an overview of the remaining three 'shrink-wrapped' documents published in 2013 by the BIM Task Group, i.e. the BIM protocol, the Scope of Services and the PII guidance documents. The material in this chapter is deliberately kept brief, and only the key issues of practical relevance have been included.

6.2 BIM protocol

The primary objective of the protocol is to enable the production of building information models at defined stages of a project. The protocol is aligned with the UK Government's BIM strategy, and incorporates provisions that support the production of deliverables for 'data drops' at defined project stages. The protocol also provides for the appointment of an 'information manager'.

A further objective of the protocol is that its use will support the adoption of effective collaborative working practices in project teams.

The BIM protocol has been divided into the following sections:

- general principles
- how the protocol works
- information manager
- information requirements
- MPDT (Model Production and Delivery Table)
- definitions:
  - priority of contract documents
  - obligations of the employer
  - obligations of project team members
  - electronic data exchange
  - use of models
  - liabilities in respect of a model
  - termination

- Appendix 1 – MPDT
- Appendix 2 – generic framework for information requirements

The following principles have formed the drafting of the protocol (BIM Protocol, 2013):

- the protocol makes the minimum changes necessary to the pre-existing contractual arrangements on construction projects
- the protocol ensures that there is an obligation on parties to provide defined elements of their works/services using models
- the protocol is a contractual document that takes precedence over existing agreements
- the protocol is flexible and should be suitable for use on all Level 2 BIM projects

The BIM protocol lays down that the protocol should form a part of all agreements or contracts between the employer and the project team members. It outlines in very comprehensive terms how it is to be incorporated into different types of contracts based on various procurement routes. For example, in projects with separate appointments, it suggests that the protocol should be appended to the appointments of members of the design team and to the building contract (BIM Protocol, 2013). There are other examples given for different procurement routes, such as design and build, and others, and the interested reader is directed to the BIM protocol document. However, it is sufficient to say that the BIM protocol is to be incorporated into all kinds of contracts, regardless of the procurement route used.

As mentioned above, the protocol document deals with various issues; all will not be considered here, only the key areas that are of relevance from a practical perspective. The next area of great importance that the protocol deals with is that of data ownership and usage rights. It is well known that this is an area of huge concern for most stakeholders in the industry and is one of the negatives (as they perceive it) about BIM adoption. The protocol uses a general concept of 'Permitted Purpose' to define the *licensed* uses of models, rather than stating the specific uses of each model. The permitted purpose is defined as: 'a purpose related to the Project (or the construction, operation and maintenance of the Project) which is consistent with the applicable Level of Detail of the relevant Model (including a Model forming part of a Federated Model) and the purpose for which the relevant Model was prepared' (BIM Protocol, 2013). The document then goes into great detail of what the liabilities of the employer and the project team members are in different circumstances.

Finally, the protocol lays down the requirement that the employer is responsible for appointing the information manager. The industry is at a turning point in relation to the role of information manager, and there is a great deal of confusion as to who is the most appropriate and best placed person to take on this role. Some feel that the design lead is the most appropriate person to take on this role, whilst others feel that it should be the project lead. Regardless of who actually performs this role, the protocol requires the employer to appoint someone. This requirement may well create a new role within employer organisations. Arguably, someone from the employer's team is indeed best placed to not only set the information requirements (EIRs), but to also manage the information delivery cycle throughout the project and handover.

The BIM protocol provides two very useful appendices at the end of the document. Appendix 1 gives a specimen MPDT (reproduced here as figure 6.1 here for convenience), which is an essential element of the agreement between all parties that lays down in very precise terms the information contained within a model at different stages of a project and information delivery data drop points. Models need to meet the LOD and other criteria at the project stages or data drops stated in the table.

The MPDT is a key element as it both allocates responsibility for preparation of the models and identifies the LOD (Level of Detail) that the model should comply with at each stage of the project and data drop points. These LODs are based on the ones elaborated in PAS 1192:Part 2. However, section 4.5.7 has provided a comprehensive account on this earlier.

Finally, a framework for a generic information requirement has been included in Appendix 2 to the protocol. Admittedly, it is a fairly short one, and there are various other additional sections for information requirement documents that would be necessary in a more realistic version of this. However, this is a useful starting point.

6.3 Outline Scope of Services for information management

The two shortest documents published by CIC in the '*shrink-wrapped*' suite are the Scope of Services and PII guidance. This section gives a brief overview of these two documents. As mentioned previously, the most important documents are the EIR, BIM protocol and PAS 1192:Parts 2 and 3[1]. These are the ones that would be needed to be consulted more than any other document in the suite that the CIC have published so far. However, an overview of these other documents is included here for completeness.

The Outline Scope of Services for information management gives high-level guidance for the kinds of activities that such a role would entail on all projects. This document provides a list of typical activities that should be included in the service provider appointments so they know exactly what they need to be responsible for in the project insofar as information management is concerned.

The role description specified in this Outline Scope of Services is based on the general obligations defined in the CIC report on the Scope of Services for the role of information management (CIC, 2013c). The CIC Scope of Services comprises lists of tasks that are, or may be, required on all projects. From these lists, all project stakeholders can draw up schedules of services for the appointment of consultants, specialists and contractors by allocating the tasks to whoever is to undertake them. The Outline Scope of Services document deals with:

1 BS 1192:Part 4 has also been (October 2014) published, which deals with COBie standards.

Specimen Model Production and Delivery Table

Showing models required at different project stages

LOD definitions (from PAS 1192)

1 Brief
2 Concept
3 Developed Design
4 Production
5 Installation
6 As constructed
7 In use

Stage definitions (from APM)

0 Strategy
1 Brief
2 Concept
3 Definition
4 Design (production information)
5 Build & Commission
6 Handover & Closeout
7 Operation and end of life

Model Originators identified by name

	Drop 1 Stage 1		Drop 2a Stage 2		Drop 2b Stage 2		Drop 3 Stage 3		Drop 4 Stage 6	
	Model Originator	Level of Detail	Model Originator	Level of Detail	Model Originator	Level of Detail	Model Originator	Level of Detail	Model Originator	Level of Detail
Overall form and content										
Space planning	Architect	1	Architect	2	Contractor	2	Contractor	3	Contractor	6
Site and context	Architect	1	Architect	2	Contractor	2	Contractor	3	Contractor	6
Surveys							Contractor	3		
External form and appearance			Architect	2	Contractor	2	Contractor	3	Contractor	6
Building and site sections					Contractor	2	Contractor	3	Contractor	6
Internal layouts					Contractor	2	Contractor	3	Contractor	6
Design strategies										
Fire			Architect	2	Contractor	2	Contractor	3	Contractor	6
Physical security			Architect	2	Contractor	2	Contractor	3	Contractor	6
Disabled access			Architect	2	Contractor	2	Contractor	3	Contractor	6
Maintenance access			Architect	2	Contractor	2	Contractor	3	Contractor	6
BREEAM					Contractor	2	Contractor	3	Contractor	6
Performance										
Building	Architect	1	Architect	2	Contractor	2	Contractor	3		
Structural	Architect	1	Str Eng	2	Contractor	2	Contractor	3		
MEP systems	Architect	1	MEP Eng	2	Contractor	2	Contractor	3		
Regulation compliance analysis							Contractor	3	Contractor	6
Thermal Simulation							Contractor	3	Contractor	6
Sustainability Analysis							Contractor	3	Contractor	6
Acoustic analysis							Contractor	3	Contractor	6
4D Programming Analysis										
5D Cost Analysis										
Services Commissioning							Contractor	3	Contractor	6
Elements, materials components										
Building			Architect	2	Contractor	2	Contractor	3	Contractor	6
Specifications			MEP Eng	2	Contractor	2	Contractor	3	Contractor	6
MEP systems					Contractor	2	Contractor	3	Contractor	6
Construction proposals										
Phasing							Contractor	3		
Site access							Contractor	3		
Site set-up							Contractor	3		
Health and safety										
Design							Contractor	3		
Construction							Contractor	3		
Operation							Contractor	3	Contractor	6

Figure 6.1: Specimen MPDT (reproduced from BIM Protocol, 2013)

- CDE management
- collaborative working, information exchange and project team management
- project information management
- other additional services

This document essentially lays down the kinds of activities that need to be undertaken for the role of information management. This role can be performed by anyone in the project team, but does not entail any design responsibility. Therefore, anyone with design responsibility may take on this role and equally someone else without a design role (e.g. the main contractor) could also take it on. As pointed out earlier, according to the BIM protocol, the appointment of someone to this role is the employer's responsibility. It should also be pointed out that the BIM Task Group (www.bimtaskgroup.org) indicates that two versions of the Scope of Services are to be used. One of these is the Outline Scope of Service for the role of information management (CIC, 2013c), which is already published. However, the other one, called the Co-ordinated Scope of Services for the role of information management is still to be published.

6.4 Best practice guide for PII for BIM models

As far as BIM Level 2 is concerned, one of the guiding principles for the Government has always been to eliminate any major disruptions to the way that the industry works. For example, a federated model structure in Level 2 means that the ownership of the individual models still stays with the parties that create the models. What this accomplishes from a contractual perspective is of enormous significance. The PII guidance document essentially articulates that Level 2 BIM should not give rise to any major insurance issues because of this federated model structure. It is hoped that Level 2 BIM will be a major step towards cost efficiencies through higher productivity. It is seen to be a major milestone en route to the final destination of Level 3 BIM, where a fully integrated model is to be used with shared ownerships (and consequently shared liabilities). This is clearly some while away, not least because of major contractual and insurance issues. Therefore, as far as Level 2 BIM is concerned, if one maintains the status quo vis-à-vis PII, it is unlikely that the PII will be compromised in any way.

6.5 Summary

This chapter presented an overview of three key publications of the BIM Task Group, viz. the BIM protocol, Scope of Services and the Professional Indemnity Insurance guidance. It was stressed how the BIM Protocol, in particular, should be very comprehensively developed for every project and should be added as a wrapper around existing contracts. This is the document that contractually binds all stakeholders to production, storage, usage and ownership of models produced during the course of a project. Any specific peculiarities of one's own organisation's workflows and processes and their impact on these issues should be reflected in the BIM protocol document as comprehensively as possible. The Scope of Services document was reviewed to outline the appointments and roles around information management in a project. Finally, the professional indemnity insurance guidance document was reviewed briefly to point out the key issues of model production and usage in relation to insurance cover one may require to be in place.

7

Training and education

7.1 Introduction

In order to get up to speed with all the key ingredients of BIM adoption, there are different kinds of training required, not just technical software training. As mentioned in the introductory chapter, BIM is about much more than technology. If anything, it is a project delivery or procurement route. In addition, as BIM requires a process-level change, the biggest training and education required is that of the mindset. Therefore, there are issues that personnel at the highest levels of an organisation need to understand and appreciate, in addition to the middle management right down to the lowest ranks of workers. In light of this, the BIM Task Group's training and education subgroup have developed an LoF (Learning Outcomes Framework), which is discussed in the next section.

The Task Group has carried out research on who the main training providers are, as well as the affected assets, roles, life-cycle stages and decision-making types for BIM adoption in the construction industry. Figure 7.1 (BIM Task Group, 2012) summarises the findings. It is a bubble diagram, with the sizes of the bubbles indicating the relative size of the concerned item in the diagram.

7.2 LoF

The BIM Task Group subsequently published a Learning Outcomes framework (LoF)

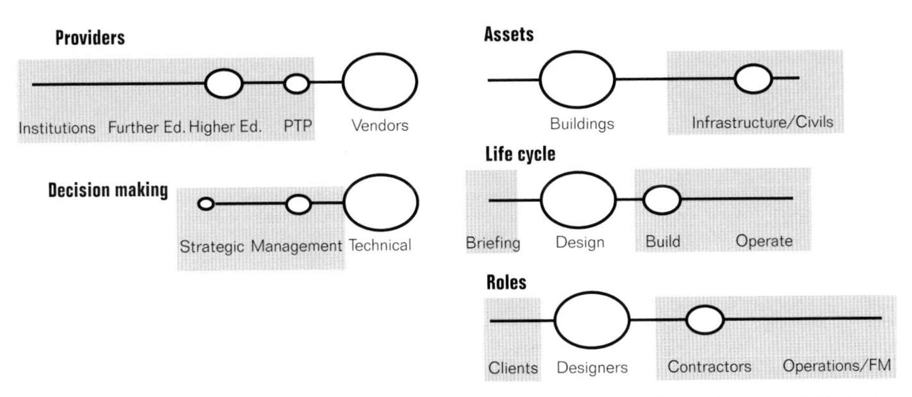

Figure 7.1: Findings of a survey on training issues (redrawn with permission from the BIM Task Group)

(BIM Task Group, 2012) to guide organisations in their effort to develop a training strategy for their staff in relation to bringing their expertise to a level commensurate with Level 2 BIM implementation in projects. The LoF is meant to be used to help structure the training strategy for Level 2 BIM and suggests structuring any training around three distinct categories (BIM Task Group, 2012):

- Strategic – this includes training on issues that concern the organisation at the highest levels and should be directed towards the top management. The materials developed for this category of training should address the benefits that the organisation may accrue as a whole in achieving all the key parameters of success that they may be working with.
- Management – this should include training the middle management, both at the organisation and at the project levels. Under this category, the key aspects of processes and standards that include contractual aspects should be covered.
- Technical – this category should focus on training related to technological aspects of BIM implementation.

Clearly this classification has implications on the kind of personnel who should be trained under the different categories. For each category, a very comprehensive set of learning outcomes have been developed by the Task Group, and can be downloaded from BIM Task Group website (BIM Task Group, 2012).

7.3 Critique of the LoF

These are clearly very comprehensive sets of outcomes. However, the way these have been worded (in some cases) may make it difficult to design a training course content that would map onto these easily. For example, 'Apply information technology to projects' appears quite frequently in several places. As this is quite a high-level and generic outcome, how does one go about mapping on to this outcome without being a great deal more specific?

Some outcomes in the LoF may be circular in that they may already be subsumed within others. For example, Tech 12 (please refer to Appendix 4 or BIM Task Group (2012)), appears to be one. How could one 'understand how to gather, manage and use BIM data…' without first actually having an understanding of 'managing and operating technical information systems'.

Under the technical outcomes, generally these are still abstractions of the traditional 'silo-based' way of working. It is not clear that the outcomes enshrined show how BIM technologies could facilitate a collaborative way of doing things. None of the outcomes actually spell that out in any way. The main aspiration behind applying BIM in projects is that the data and information are gradually evolved to a full set at handover. This gradual evolution of data generation, gathering, storage and use is a key element that needs to be reflected in these outcomes somehow. In fact, the way COBie sheets are meant to be filled in gradually from inception to handover (COBie, 2012) is probably a good way of demonstrating this point.

Somewhere in the outcomes, arguably, there should also be some related to monitoring and assessing the outcomes of a project in relation to BIM in value terms. So, an outcome on the lines of '... should be able to assess the value generated by using BIM on the project' may be useful. This will also close the loop in terms of tying back to some of the other strategic outcomes.

Continuing on from the last point, the management and technical outcomes should tie in with strategic outcomes because these are lower level abstractions of the strategic outcomes and should, in theory, map onto each other quite well. So, there should be a link provided from a technical outcome back to a strategic one, suggesting how one would be achieving a particular strategic outcome by achieving a management and/or a technical one.

7.4 Suggested training framework for BIM

In order to ensure that an organisation gets the required levels of expertise for using BIM technologies and processes on all projects, one has to essentially have a BIM strategy for the organisation. As shown in figure 7.2, there are typically three elements to this. It is recommended that the first thing an organisation needs to have in place is a BIM champion. This has to be a person with a passion for BIM and should be sufficiently knowledgeable in key aspects and issues of BIM. Once the champion is in place, one of the key activities is to identify the main challenges that the organisation might be facing vis-à-vis BIM adoption. This may range from the usual resistance to change to lack of infrastructure and even to a lack of sufficient expertise within the organisation. It is not sufficient to identify the internal challenges; the external challenges that one might be facing also need to be identified. This might range from stiff competition from other firms to shortage of skills available locally. It is only after a proper grasp of these issues has been developed that one is in a position to develop a training strategy and policy for the organisation.

It must be pointed out that although the BIM Task Group's LoF categorises training into three groups, it does not actually break down some key aspects of the different kinds of training required. For example, when it comes to training related to cultural aspects such as collaborative ways of working, the LoF does not cover these

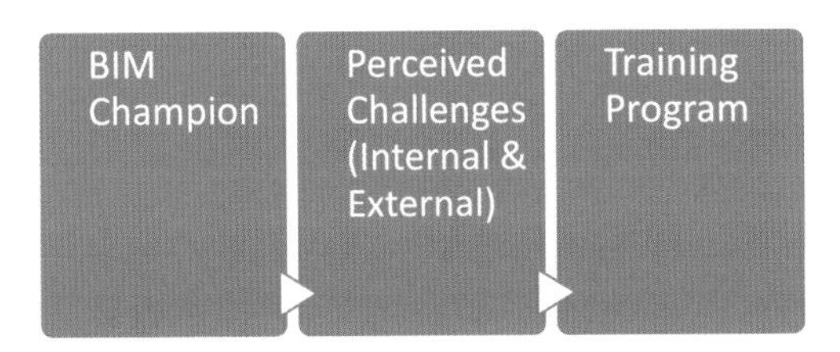

Figure 7.2: Key aspects of BIM training strategy

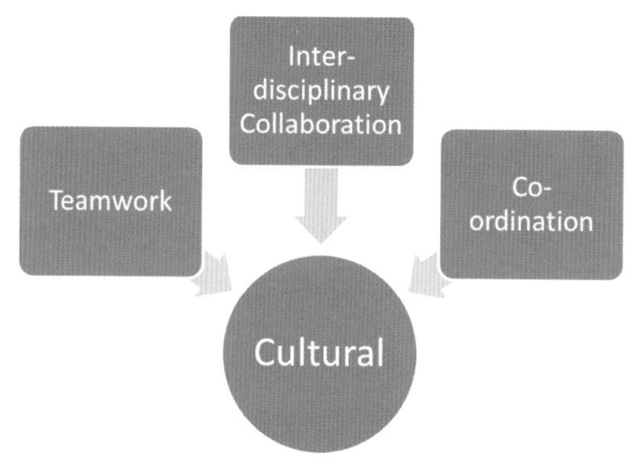

Figure 7.3: Key cultural and softer aspects of BIM training strategy

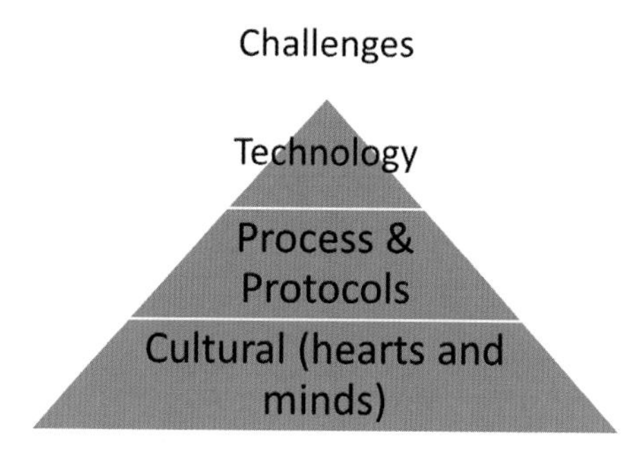

Figure 7.4: Key challenges in developing BIM training strategy

issues at all. Some of these aspects are shown in figure 7.3. It should be noted that obtaining technical training in software skills is easier than all the other aspects that BIM adoption requires. In fact, it is also a matter of common knowledge that simply going through a training course is never enough to ensure that the real benefits of any training accrue to the organisation. Any training has to be followed up by substantial hands-on implementation of topics and issues covered in real scenarios. Figure 7.4 illustrates some of the key challenges likely to be faced by any organisation in order of importance and complexity as a pyramid.

7.5 Suggested road map for BIM training

The following gives a suggested road map for BIM training:

- knowledge dissemination of the Government's policy on BIM objectives, key drivers and background information
- information dissemination on the 'shrink-wrapped' documents published by the BIM Task Group

- organisational training needs assessment and analysis
- development of a comprehensive training plan
    - categorisation of all training into the categories mentioned in the LoF
    - identification of personnel and their mappings into the different categories
    - identification of each relevant individual's training needs
    - scheduling of the specific training programmes
    - execution
    - feedback and monitoring
- revision of the training strategy and policy in light of feedback and monitoring of earlier stages

7.6 Nature of organisation

The nature of an organisation's main activities will also have a profound impact on the kind of training required. For example, an employer organisation needs a particular emphasis and slant in its training strategy as opposed to a contracting or a design organisation.

An employer (client) organisation needs to focus more on issues such as EIRs and information management than perhaps specific modelling technologies. Having said that, there will be overlaps between requirements of all kinds of organisations in some aspects, for example an understanding of a standard data format to be used for information exchange (such as the COBie UK 2012 dataset), general BIM protocols, an appreciation of contractual and insurance implications of Level 2 BIM, etc.

7.7 Example training strategy road map for an employer/ client organisation

This section provides suggestions for a road map for developing a training strategy for a typical client organisation:

- Map out the key processes from a project/asset need to procurement to later stages of design, construction, handover and O&M.
- Map the internal process onto the RIBA PoW 2013, OGC Gateways and finally on to BIM processes as mentioned in chapter 5 of PAS 1192:Part 2. These BIM processes will typically consist of development of EIRs, development of BIM PEP, collation of BIM protocols, data drop point linked information exchanges with the final output of AIM aligned to GSL principles, export of AIM data seamlessly to a CAFM system at handover and processes for post-occupancy evaluation data capture.

- Develop for each stage in the BIM processes the LOD and Level of Detail that the models should capture.
- Develop the PLQs for each stage of the BIM processes developed above. These PLQs will be used to assess whether the project can proceed to the next stage in the process at every key stage.
- Assess the BIM capability for each stage in the BIM processes outlined above.
- Develop detailed training needs and plans for each category of strategic, management and technical training.

It should be pointed out that without first mapping out the processes that the organisation actually follows in key activities, it would be futile to arbitrarily jump to a training strategy. Therefore, it is recommended that in case there are no formal processes that the organisation follows in procurement of assets as a client organisation, the first step is to develop one before identifying a BIM strategy and the training required for bringing the organisation up to speed with Level 2 BIM. It should be clear that the most important issue for a client organisation is the ability to develop a robust set of information requirements and to monitor at each stage whether its entire supply chain is complying with those requirements. Obviously, the associated issues of standards and protocols for information exchange at every stage also need to be assessed. Therefore, in this sense, the client organisation does not need to be as adept at modelling and technological aspects of BIM as designers and contractors would need to be. So, if a contracting organisation was to develop its BIM strategy, the focus and steer needed for them would be markedly different. As a starting point, the guidance notes given on Level 2 BIM compliance in Appendix 1 are relevant for a design and contracting organisation and can be used to develop a training strategy for such organisations.

7.8 Challenges in implementing a training programme

The preceding sections have provided background material for developing a road map for a training strategy. However, there are several challenges normally encountered in most organisations in achieving success in this endeavour. Some of these challenges are:

- lack of formal processes for key activities of the organisation
- lack of top management buy-in
- insufficient training needs analysis
- lack of a BIM champion
- lack of a formal training plan
- follow-up post-training
- monitoring of effectiveness post-training
- lack of a feedback loop for continuous improvement of the current training strategy

Some sample training course topics and associated timelines and topics are given in Appendix 4. These are based on the BIM Task Group's LoF, and are provided here as suggested training courses and are not definitive by any means. Each organisation will need to assess its own requirements based on some of the criteria mentioned above before developing its own training programme.

7.9 BIM Academic Forum

The BAF (BIM Academic Forum) was formed in 2012 to bring together and exchange ideas between academic institutions for incorporating BIM into their curricula. The BAF developed its own BIM curriculum with learning outcomes, which is based on the BIM Task Group's LoF. The BAF's curriculum is meant for academic curricula at different levels of higher education programmes. However, these could be helpful for some organisations in developing their own training programmes. The BAF's curriculum can be downloaded from the HESA website (Higher Education Statistics Agency, www.hesa.ac.uk) (BAF, 2013).

7.10 Summary

This chapter has presented a brief outline of training issues related to BIM adoption by different kinds of organisations. The background to the LoF published by the BIM Task Group was also covered, including a brief critique of the framework. The nature of an organisation's activities and its impact on training strategy in relation to the BIM was presented, and a framework for developing a training strategy along with the main potential challenges was also outlined.

References

AEC Magazine, *BIM: What Your Government Wants,* AEC Magazine, 28 January 2012.

AIA (2013), *Level of Development Specification*, American Institute of Architects, 2013.

APM, *APM Body of Knowledge*, http://www.staffs.ac.uk/sgc1/faculty/business-project-management/documents/APM_BoK_000.pdf, 2014.

BAF, *Embedding Building Information Modelling (BIM) within the Taught Curriculum*, www.heacademy.ac.uk, 2013.

Belsky, M., Sacks, R. and Brilakis, I., A Framework for Semantic Enrichment of IFC Building Models. *Proceedings of the 30th CIB W78 International Conference*, 9 12 October 2013, Beijing, China

Bew, M., *BIM Investor's Report*, IGI Global, 2010.

Bew, M. and Richards, M., *BIM Maturity Levels*, 2008 (as referenced on www.bimtaskgroup.org).

BIM Forum (2013), *Level of Development Specifications for Building Information Models*, www.bimforum.org/lod, August 2013.

BIM Protocol, *The BIM Protocol*, Construction Industry Council, 2013.

BIM Task Group, *Learning Outcomes Framework*, http://www.bimtaskgroup.org/education-and-training, 2012.

Bronowski, J., *The Origins of Knowledge and Imagination*, Yale University Press, 1978.

BSI (2008), BS 1192:2007, Collaborative Production of Architectural, Engineering and Construction Information. Code of Practice, British Standards Institution, January 2008.

BSI (2014), BS 1192-4:2014, Collaborative Production of Information. Fulfilling employer's information exchange requirements using COBie - Code of Practice, British Standards Institution, September 2014.

BSI (2010), Building Information Management- A Standard Framework and Guide to BS 1192, BSI, 2010.

http://www.buildingsmart.org/standards/ifc/model-industry-foundation-classes-ifc.

Cheng, J.C.P., Zhang, Y., Han, C.S., Law, K.H. and Kumar, B., Web-enabled Model-based CAD for Design, *ICCCBE Conference*, Nottingham University, 2010.

CIC (2013a), *Employer's Information Requirements*, Construction Industry Council, February 2013.

CIC (2013b), *PAS 1192:Part 2*, Construction Industry Council, February 2013.

CIC (2013c), *Scope of Services*, Construction Industry Council, 2013.

CIC,(2013d) *Best Practice Guide for Professional Indemnity Insurance when using Building Information Models*, Construction Industry Council, 2013.

CIC (2014), *PAS 1192:Part 3*, Construction Industry Council, 2014.

CIOB (2002), *Code of Practice for Project Management for Construction and Development*, Blackwell Publishing, 2002.

COBie, *Construction Operations Building Information Exchange*, BuildingSmart, www.wbdg.com/resources/cobie.php/, 2012.

Construction 2025, Her Majesty's Stationery Office, July 2013.

CPIC, *Pre and Post-Contract BEP*, Construction Project Information Committee, 2013.

Crotty, R., *The Impact of Building Information Modelling*, Spon Press, 2012.

Digikey, http://www.digikey.com/en/articles/techzone/2011/mar/intel-increases-its-microprocessor-market-share, 2011.

DPoW, *Digital Plan of Work*. BIM Task Group website, www.bimtaskgroup.org, March 2013.

Egan, J., *Rethinking Construction*, Her Majesty's Stationery Office, 1998.

Eastman, C., Teicholz, P., Sacks, R. and Liston, K., *BIM Handbook*, 2nd edn, Wiley, 2012.

Elmasri, R. and Navathe, S.B., *Chapter 3, Fundamental of Database Systems*, 2nd edn, Addison-Wesley, pp. 39–68, 1994.

Fruchter, R., Schrotenboer, T. and Luth, G.P., From Building Information Model to Building Knowledge Model, *Proceedings of ASCE Computing in CE Conference*, pp. 380–389, 2009.

Gallaher, M.P., O'Connor, A.C., Dettbarn, J.L. and Gilday, L.T., *Cost Analysis of Inadequate Interoperability in the U.S. Capital Facilities Industry*, National Institute of Standards and Technology, Gaithersburg, Maryland, 2004.

Government Construction Strategy, Her Majesty's Stationery Office, May 2011.

Kennedy, P., Milligan, J., Cattanach, L. and McCluskey, E., The Development of Statutory Adjudication in the UK and its Relationship with Construction Workload, *Adjudication Trends COBRA Conference*, 2010.

Kennerson, Patrick, Master's Project Management Guest Lecture, School of Engineering and Built Environment, 2013.

Khemlani, L., *The IFC Model: A Look Under the Hood*, AECBytes Features4, www.aecbytes.com, 2004.

Latham, M., *Constructing the Team*, Her Majesty's Stationery Office, 1994.

Lipman, R.R., *CIS/2 and IFC for Structural Steel*, Building Smart Day, Washington D.C., November 2006.

MIT BIM PEP, Department of Facilities, Massachusetts Institute of Technology, Massachusetts, USA, 2014.

Murray, M. and Langford, D., *Construction Report, 1944–1990*, Blackwell Science, Oxford, 2003.

NBIMS National BIM Standard - United States. National Building Information Model Standard Project Committee, http://www.nationalbimstandard.org/faq.php#faq1, accessed 20 November 2013.

ONS (2009), *Construction Statistics Annual*, Office of National Statistics, 2009

PSU BIM PEP, Computer Integrated Construction Research Group, Pennsylvania State University, USA, 2010.

RIBA, 2013. RIBA Plan of Work, http://www.ribaplanofwork.com/, 2013.

Scottish Construction Procurement Report, Scottish Government, October 2013.

Sommerville, J. and Craig, N., Sysdox as a Decision Support Tool, *International Conference on Decision Making in Urban and Civil Engineering*, London, December 2002.

Uniformat (2010), UniFormat Master Guide Specifications, http://www.csinet.org/Home-Page-Category/Formats/Master-Guide-Specs

Voss, E. and Overend, M., A Tool that Combines Building Information Modeling and Knowledge Based Engineering to Assess Façade Manufacturability, *Advanced Building Skins*, 2012.

Appendix 1

Brief guide to Level 2 BIM compliance/capability

One of the issues that many firms appear to be agonising over in the run up to 2016 is how they can demonstrate Level 2 competence and capability to convince prospective clients. This appendix attempts to answer some of the key questions that are generally being asked at the PQQ stage by several client organisations. A real questionnaire from a client organisation is included in Appendix 2.

In addition, some other key questions that many organisations are carefully considering are given below:

What is the best BIM software we should buy?

This is not a question that deserves a great deal of thought. The reason being is that the answer to this question depends on a few simple factors. It is not the intention here to recommend any particular piece of commercial software, but to give some general guidance as to what characteristics to look for when making a decision.

- What kind of organisation are you?

If you are an architectural firm, you clearly need a good BIM authoring tool. There are several good BIM tools available on the market, but as far as the *best* one is concerned, it is very difficult to compare them from *your* perspective. Having said that, there are comparisons readily available from various sources. For organisations not involved in design or costing or scheduling and project management, one may not even need to buy any BIM tool. This is an important issue that many companies do not seem to appreciate. Many organisations in a typical supply chain (e.g. window and door suppliers or fitters) could do with just a building information model viewer, which can be obtained free of charge on the Internet.

- Is the software under consideration IFC-compliant?

This should be a key criterion, as the ability to interoperate with other software is a key factor in BIM-enabled projects. As mentioned in chapter four, even buying IFC-compliant software does not guarantee completely problem-free interoperability with other tools, but this is the best one can do.

- What is the budget?

This is self-explanatory. However, the important point in this regard is that you need not go overboard with spending over the odds. This is because as long as *your* software can interoperate with other tools and it can do what you need to do with it, it does not matter if it is the slickest or the market-leading software or not.

- We now have several BIM software licences and have gone through training our technicians, is there anything else we need to do?

Having gone through a book like this, it is hoped that the answer to this question is self-explanatory. However, this question has been included here simply to re-emphasize that just having BIM software licences is nowhere near enough to position any organisation for 2016. Here is a list of key issues that every organisation should seek to address in order to position itself for 2016:

- BIM/information management policy
- BIM processes:
    - awareness and understanding of key processes, standards and protocols
    - awareness and understanding of 'shrink-wrapped' CIC/ BIM Task Group publications
    - education and training for information management
    - technology and infrastructure
- Uniclass awareness and usage track record
- PE usage and awareness and track record

Appendix 2

Typical responses to a BIM-based PQQ

This appendix contains a typical PQQ being used by several client organisations. It is included here to provide guidance on how to respond to such PQQs.

Many clients will exempt someone from this PQQ if they hold some kind of third or external party certificate of compliance with PAS 1192:Part 2 from any organisation with an accreditation awarding authority or equivalent. Until recently, such accreditations were not even available. However, some organisations have recently started providing certificates of compliance. Having said that, one should be careful to not turn this into just a paper exercise and focus more on developing real skills and expertise, which can only be achieved by going through a structured training plan and then implementing the skills in some pilot projects.

Do you hold an assessment certificate of compliance with BIM Level 2?

YES/NO

(If no, see questions below that are required to be answered)

Do you have the expertise for working on a project implementing the information delivery life cycle as described in PAS 1192:2:2013?

YES/NO

(If yes, details must be provided)

Suggested response:

We have a BIM champion (name?) and a group within the organisation tasked with establishing the overall BIM strategy, development of BIM standards and processes as well as training programmes for the whole organisation. The group has close links with the GCU (Glasgow Caledonian University) BIM Centre (*or anyone else as appropriate*). The GCU BIM Centre has a strong relationship with key members of the UK Government BIM Task Group and is therefore fully aware of standards and other guidance documents published by the BIM Task Group. We feel confident that, through these links and the effort of our BIM Group, we are in a strong position to stay abreast of the key developments and to embed these in our organisation relatively quickly. We can confirm that we understand the concepts behind CDE as described in PAS 1192:2:2013. Although we have not had the opportunity to work with a CDE as envisaged in PAS 1192, we do have experience with similar environments, most notably with 4Projects (*or any other that the organisation may have used before*). We have worked with 4Projects on several projects – (project details, supply chain, maybe some statistics on numbers or kinds of documents exchanged through the environment, etc.). We do appreciate that 4Projects is fundamentally a PE environment, but unlike several other PE environments, it comes closest to the kind of environment that CDE is purported to be. We have discussed this with key members of the Task Group actively involved in the drafting of PAS 1192, and they are generally in agreement with our view that the main difference between a typical PE environment and CDE is that CDE is a more actively managed version of a PE with the incorporation of WIP and gateways for quality control of what is 'dropped' into the CDE.

Please demonstrate that your organisation has documented policies, systems and procedures to achieve 'Level 2 BIM' maturity as defined in the Government's BIM strategy?

YES/NO

(If yes, please provide details)

Describe the WIP, perhaps with some specific examples.

We have a good understanding of the guidance documents published by the Task Group and CIC at the end of February 2013 onwards. In particular, we are developing processes and policies in line with PAS 1192:2:2013. We endeavour to develop our own set of policies and procedures based on these documents and the Government's BIM strategy, where possible, and seek to refer to other sources of information for the areas where we still await guidance. An example is the definition of LOD, which has to be specified by the employer but as stated in the EIR document (p. 5, reference 1.1.4), 'In advance of publication of definitive UK definitions of Levels of Detail for models, assemblies and components, employers can use existing generic definitions such as that included within PAS 1192-2 as a reference point'. PAS 1192:Part 2 (Appendix A.78), on the other hand, points to the CIC Scope of Services document for a definition of LOD. The trail ends on the Task Group website, which states that the detailed Scope of Services document is still under preparation. This highlights the fact that there are a number of areas where significant gaps exist, and our approach is to draw upon other sources to develop our own policies in such areas (e.g. for LOD, AIA document E202 is a reference point that we are using). We are making extensive use of frequently asked questions and PLQs, as well as other resources in the Task Group Labs portal on the BIM Task Group website. However, in several areas (e.g. LOD), we will work with the information provided to us in the EIR documents and agree a mutually workable set of policies and procedures with the employer.

Demonstrate your capability in developing or already having a standard BIM PEP (or BEP) template that could be adapted to specifics of particular projects.

YES/NO

(If yes, please provide details)

Suggested response:

We are fully aware of the BEP documents as described in PAS 1192:2:2013. We are also aware of various other BEPs from other UK as well as overseas organisations (e.g. CPIC BEP, PSU BIM PEP) and are currently in the process of developing our own BEP document both for pre- and post-award scenarios. As far as COBie is concerned, we are developing an understanding of the COBie UK 2012 dataset and are familiar with its requirements. However, we have not had the opportunity to work with COBie on any project so far. Through our existing relationships and partnerships with organisations with BIM expertise (e.g. GCU), we feel confident that we have direct access to the latest developments in all these areas and are in a position to incorporate any relevant material into our own BIM processes and documentation quite quickly.

Do you have a training strategy and plan for your staff in BIM skills?

YES/NO

(If yes, please provide details)

Please demonstrate that your organisation has in place training arrangements to ensure that your staff/workforce have sufficient skills and understanding to implement and deliver projects in accordance with the policy and procedures established to achieve 'Level 2 BIM' maturity.

This section should address the following key points:

- Existing (if any) relationships with other organisations with BIM expertise. This is particularly important when internal expertise in BIM does not quite exist yet.
- Awareness and understanding of the BIM Task Group's LoF in terms of categorisation of training activities into strategic, management and technical.
- Knowledge of the BIM Task Group's training-related activities for early adopter projects within the Ministry of Justice and other Government departments.
- Some statistics (if possible) on staff who have already attended CPD courses and technology-based training on BIM software, etc.
- Awareness of CPIx templates. Perhaps simply state that, 'we are aware of these templates and are in the process of adapting these to our own specific needs'.
- Information on existing internal training programme/plan monitoring processes.
- An outline of the training programmes the company might be likely to introduce as BIM processes are developed.

Appendix 3

Sample BIM PEP for a building services company

Acknowledgement is given here to B. Kilpatrick and J. Vincent of Hulley & Kirkwood for providing permission to reproduce this BIM PEP.

HULLEY & KIRKWOOD CONSULTING ENGINEERS LIMITED

BIM PROJECT EXECUTION PLAN

FOR

[PROJECT TITLE]

[Date]

Hulley & Kirkwood Consulting Engineers Ltd
Head Office
Watermark Business Park
305 Govan Road
Glasgow
G51 2SE

(t): 0141 332 5466
(f): 0870 928 1028
(e): hk.glasgow@hulley.co.uk
(w): www.hulley.co.uk

Prepared by:
Authorised by:
Revision:
Date:
File location:

REV	DESCRIPTION	PREPARED BY	DATE
Issue no. 1	First issue	XXXXX	XXXXX

Table of contents

Section A: BIM project execution planning guide overview

To successfully implement BIM on a project, Hulley & Kirkwood has developed this detailed BIM PEP. The BIM PEP defines uses for BIM on the project (e.g. design authoring, design reviews, 3D co-ordination and as-built record modelling), along with a detailed process for executing BIM on this project.

The enclosed document is only intended to relate to the Hulley & Kirkwood/ MEP element of the project. A comprehensive PEP will need to be created for the project that encompasses all participants in the project.

Who the lead is in creating the comprehensive PEP needs to be discussed and agreed, but it is unlikely to be Hulley & Kirkwood. However, the enclosed document will enable our element to be readily incorporated into the overall PEP as well as identifying the level and amount of information that will be incorporated in our building information model.

The document is a living document and will need to be reviewed and updated as changes to the project occur or additional parties come on board as the project progresses.

To assist in the completion of this document, reference should be made to the 'BIM PEP Guidance' document that can be found on the Hulley & Kirkwood knowledge management system. This document should be read either in conjunction with or in advance of completing the enclosed PEP document.

Section B: Project information (to be completed by H&K)

1. CLIENT:	
2. PROJECT NAME:	
3. PROJECT NUMBER:	
4. CONTRACT TYPE:	
5. BRIEF PROJECT DESCRIPTION:	

6. ADDITIONAL PROJECT INFORMATION:	
7. HULLEY & KIRKWOOD ROLE/ DUTIES:	
8. SUMMARY OF HIGH-LEVEL BIM INPUT/DELIVERABLES EXPECTED FROM HULLEY & KIRKWOOD (e.g. extent of modelling, amount of embedded information):	

Section C: Key project contacts (to be completed by H&K)

The following is a list of the lead BIM contacts for each organisation on the project.

ROLE	ORGANISATION	NAME	EMAIL	PHONE
Client				
Project Manager				
Architect				
Structural/Civil				
Quantity Surveyor				
CDM Coordinator				
BREEAM				
Quantity Surveyor				
Main Contractor				
Mechanical Subcontractor				
Electrical Subcontractor				

Section D: Project goals / BIM objectives (to be completed in conjunction with design team and client)

1. MAJOR BIM GOALS / OBJECTIVES:

   State BIM goals / objectives

Priority (1–3)	Goal Description	Potential BIM Uses
1. Most important	**Value-added objectives**	

2. BIM USES (to be completed in conjunction with design team and client):

   The elements that are to be included in the building information model should be identified in the table below.

OPERATE	X	CONSTRUCT	X	DESIGN	X	PLAN	X
BUILDING MAINTENANCE SCHEDULING		SITE UTILISATION PLANNING		DESIGN AUTHORING	X	PROGRAMMING	
BUILDING SYSTEM ANALYSIS		CONSTRUCTION SYSTEM DESIGN		DESIGN REVIEWS	X	SITE ANALYSIS	
ASSET MANAGEMENT		3D CO-ORDINATION	X	3D CO-ORDINATION	X		
SPACE MANAGEMENT / TRACKING		DIGITAL FABRICATION					
DISASTER PLANNING		3D CONTROL AND PLANNING		LIGHTING ANALYSIS			
RECORD MODELLING		RECORD MODELLING	X	ENERGY ANALYSIS			
				MECHANICAL ANALY-SIS			
				OTHER ENGINEERING ANALYSIS			
				BREEAM EVALUATION			
				CODE VALIDATION			
4D MODELLING		4D MODELLING		4D MODELLING		4D MODELLING	
COST ESTIMATION		COST ESTIMATION		COST ESTIMATION		COST ESTIMATION	
EXISTING CONDITIONS MODELLING		EXISTING CONDITIONS MODELLING		EXISTING CONDITIONS MODELLING		EXISTING CONDITIONS MODELLING	

Following on from the table above, an assessment of the capability of the team as a whole to provide the elements above and the value of providing that element on the project can be created using the matrix below.

BIM Use*	Value to Project	Responsible Party	Value to Responsible Party	Capability Rating	Additional Resources / Competencies Required to Implement	Notes	Proceed with Use
	High / Medium / Low		High / Medium / Low	Scale 1–3 (1 = Low) Resources Competency Experience			YES / NO / MAYBE
Maintenance Scheduling							
Building Systems Analysis							
Record Modelling							

BIM Use*	Value to Project	Responsible Party	Value to Responsible Party	Capability Rating			Additional Resources / Competencies Required to Implement	Notes	Proceed with Use
	High / Medium / Low		High / Medium / Low	Scale 1–3 (1 = Low)					YES / NO / MAYBE
				Resources	Competency	Experience			
Cost Estimation									
4D Modelling									
Site Utilisation Planning									

Layout Control and Planning			
3D Co-ordination (Construction)			
Engineering Analysis			
Site Analysis			
Design Reviews			

BIM Use*	Value to Project	Responsible Party	Value to Responsible Party	Capability Rating	Additional Resources / Competencies Required to Implement	Notes	Proceed with Use
	High / Medium / Low		High / Medium / Low	Scale 1–3 (1 = Low) Resources Competency Experience			YES / NO / MAYBE
3D Co-ordination (Design)							
Existing Conditions Modelling							
Design Authoring							

Programming

* Additional BIM uses as well as information on each use can be found at http://www.engr.psu.edu/ae/cic/bimex/.

3. BIM MODEL COMPLEXITY (to be completed by H&K):

   The below establishes the extent to which the model is being taken. The enclosed states which areas are being modelled to what level of complexity.

4. BIM MODEL EXCLUSIONS (to be completed by H&K):

   The list below identifies what elements of the final installation will be excluded from the BIM model being produced by Hulley & Kirkwood, but which will need consideration as part of the installation model to be created by the contractor/subcontractor.

Exclusions from BIM Model	
Mechanical	Pipe Coupling and Bends
	Insulation
	Ductwork Access Hatches
	Anchors
	Expansion Pieces
	Brackets
Electrical	Conduit
	Individual Cables
	Earthing and Cross Bonding
	Gas Extinguishing Pipework
	Brackets
	Containment Couplings and Bends
Specialist Design	Sprinklers (notional space only)
	BMS Containment (notional space only)
	Pneumatic Tube (notional space only)
	Gas Extinguishing (notional space only)

Section E: Organisational roles / staffing (to be completed by H&K)

The list below details those individuals from each organisation who are involved in the delivery of the uses identified in section D. This should include all contributing organisations to any one identified use.

BIM USE	ORGANISATION	LEAD CONTACT

Section F: BIM process design (to be completed in conjunction with design team and client)

1. OVERALL BIM OVERVIEW MAP:

   A typical BIM overview map is included in Appendix 2 of the Hulley & Kirkwood 'PEP Guidance' document. The map can be used as a template and should be adapted to suit the project as necessary.

2. DETAIL LEVEL BIM USE PROCESS MAP(S):

   a. Existing Conditions Modelling
   b. Cost Estimation
   c. Design Reviews
   d. Design Authoring
   e. Energy Analysis
   f. Lighting Analysis
   g. 3D Co-ordination
   h. Record Modelling
   i. Maintenance Scheduling
   j. Building System Analysis

Typical detail level process map templates exist in Appendix 2 of the Hulley & Kirkwood 'PEP Guidance' document. The maps can be used as a template and should be adapted to suit the project as necessary. Delete unused or add additional process maps to/from the list.

Section G: BIM information exchange worksheet (to be completed in conjunction with design team and client)

1. LIST OF INFORMATION EXCHANGE WORKSHEET(S):

   a. Existing Conditions Modelling
   b. Cost Estimation
   c. Design Reviews
   d. Design Authoring
   e. Energy Analysis
   f. Lighting Analysis
   g. 3D Co-ordination
   h. Record Modelling

i. Maintenance Scheduling

j. Building System Analysis

A sample information exchange worksheet is included in Appendix 3 of the Hulley & Kirkwood 'PEP Guidance' document. The worksheet can be used as a template and should be adapted to suit the project as necessary. Delete unused or add additional worksheets to/from the list.

Section H: Facility data requirements (to be completed in conjunction with client)

In the event of there being pre-determined or specific handover information required as part of the building information model, this should be documented in the Facility Data Worksheet. A copy of a typical worksheet is included in the Hulley & Kirkwood 'BIM PEP Guidance' document.

Section I: Collaboration procedures (to be completed in conjunction with design team and client)

1. COLLABORATION STRATEGY:

   Describe how the project team will collaborate. Include items such as communication methods, document management and transfer, record storage, etc.

2. MEETING PROCEDURES:

   The following are examples of meetings that should be considered.

MEETING TYPE	REQUIRED PER CONTRACT	PROJECT STAGE	FREQUENCY	PARTICIPANTS	LOCATION
BIM REQUIREMENTS KICK-OFF					
BIM EXECUTION PLAN DEMONSTRATION					

MEETING TYPE	REQUIRED PER CONTRACT	PROJECT STAGE	FREQUENCY	PARTICIPANTS	LOCATION
DESIGN CO-ORDINATION					
CONSTRUCTION OVER-THE-SHOULDER PROGRESS REVIEWS					
ANY OTHER BIM MEETINGS THAT OCCUR WITH MULTIPLE PARTIES					

Section J: Quality control (to be completed in conjunction with design team and client)

1. OVERALL STRATEGY FOR QUALITY CONTROL:

   Describe the strategy to control the quality of the model.

2. QUALITY CONTROL CHECKS:

   The following checks should be performed to assure quality.

CHECKS	DEFINITION	RESPONSIBLE PARTY	SOFTWARE PROGRAM(S)	FREQUENCY
VISUAL CHECK	Ensure there are no unintended model components and the design intent has been followed			
INTERFERENCE CHECK	Detect problems in the model where two building components clash, including soft and hard			
STANDARDS CHECK	Ensure that the BIM standard has been followed (fonts, dimensions, line styles, levels/ layers, etc.)			
MODEL INTEGRITY CHECKS	Describe the quality assurance validation process used to ensure that the project facility dataset has no undefined, incorrectly defined or duplicated elements and the reporting process on non-compliant elements and corrective action plans			

Section K: Technological infrastructure needs (to be completed in conjunction with design team and client)

1. SOFTWARE:

List software used to deliver BIM. Remove software that is not applicable.

BIM USE	USER	SOFTWARE	VERSION
DESIGN	MECHANICAL/ELECTRICAL		
DESIGN	ARCHITECT		
DESIGN	STRUCTURAL/CIVIL		
DESIGN			
DESIGN			
DESIGN			
DESIGN			
COST ESTIMATION			
EXISTING CONDITIONS MODELLING			
DIGITAL FABRICATION			
3D CO-ORDINATION			
DESIGN REVIEWS			
LIGHTING ANALYSIS			
ENERGY ANALYSIS			
BREEAM EVALUATION			

2. BIM and CAD Standards: (to be completed by H&K):

STANDARD	VERSION

Section L: Model organisation (to be completed by H&K)

1. FILE NAMING STANDARD:

   Refer to the Hulley & Kirkwood 'PEP Guidance' document.

2. MODEL STRUCTURE:

   Describe and sketch how the model is divided up. For example, by buildings, by floors, by zones, by areas and/or by disciplines.

3. CO-ORDINATE SYSTEMS:

   Describe the co-ordinate systems used.

Section M: Project deliverables (to be completed by H&K)

BIM SUBMITTAL ITEM	STAGE	FORMAT	NOTES

Appendix 4

A sample set of learning outcomes for an introductory course in Level 2 BIM

This appendix gives a list of topics and material to be covered in an introductory level training course on Level 2 BIM implementation. This is based on the BIM Task Group's LoF and can be tailored to specific requirements.

Learning outcomes for level 1 course	Topics covered
1.1 To understand the concept of BIM, within the context of process and activity aided by technology and 'data' sets and how this relates to the construction industry. To including the key interaction of process, technology and people in a BIM environment. 1 hour	• Understand the various BIM definitions • Understand the BIM maturity model and government targets (context of current industry readiness) • Understand the benefits and usage of BIM, including BIM to field, visual method statements, progress monitoring, live quality management data work flows (opportunities and reality) • Understand the requirements for structured data
1.2 To understand the principles and structure of the BSI Level 2 processes at a strategic level 45 mins	• Understand the purpose and interaction of the various process documents and artefacts

Learning outcomes for level 1 course	Topics covered
1.3 To understand the principles of digital tools, key workflows and benefits 1.5 hours	• Geometrical models 10 mins • Non-graphical data (COBie) 10 mins • Model federation and validation (e.g. clash) 20 mins • 4D simulation 10 mins • 5D cost (Capex) 10 mins • 6D life cycle / federated model 10 mins • Point clouds Including real world case study examples from Mace World
1.4 To understand the business case for BIM, taking into account investment and benefits, examining the major cost drivers and benefits of the digital information model(s)	• Understand how BIM outcomes relate to client needs • Understand how BIM can benefit supply chain (and Mace) • Opportunity for effective collaboration between the various stakeholders in the asset life cycle (CAVEs / mission rooms, etc.) • Potential use cases at operational stages • Analysis of life-cycle cost and maintenance requirements • Opportunity of other complementary digital workflows, e.g. laser scanning and data capture
1.5 To understand the process of information management and exchanges	• Understand PoWs • Understand data maturity growth / LOD • Understand data drops / exchanges
1.6 Understand commercial and legal aspects of BIM	Understand concept of model ownership Internet protocol, design responsibilities, information manager role, buying BIM as a service

Learning outcomes for level 2 course	Examples of knowledge/activities
2.1 To understand the processes and standards that are applicable to the management of information about an asset throughout its lifespan	Application of BSI 1192:2007, PAS 1192, Parts 2 and 3, BIM protocol, classifications, e.g. Uniclass 2, new rules of measurement, DPoW, EIRs and Uniclass.
2.2 To understand the principles of BS 1192:4 COBie:UK:2014	Worked example of a small COBie spreadsheet, fields, linkage with other data and classifications. Use cases.
2.3 To understand the need for the right timing of information at each stage of the project life cycle for implementation of BIM and the responsibilities of each party at these stages	Understand and review the typical information exchange requirements of project team and asset managers. Who, what and when (and why)?
Learning outcomes for level 3 course	**Examples of knowledge/activities**
3.1 To understand how to step up BIM on your project	Understand how to prepare a: • project BIM strategy • BIM EIR • BIMxP • TIDP/MIDP
3.2 To understand how to assess capabilities of the project team relative to the desired BIM outcomes	Understand how to prepare and assess BIM competency and score-carding
3.3 To understand quality management of the BIM process	Understand how to prepare a BIM quality checklist and audit the modelling / information exchange process

Learning outcomes for level 3 course	Examples of knowledge/activities
3.4 To understand how a model and CDE should be set up	• Common origin points • Model structure • System identifications • Data demand • CDE typical workflows and folder structures • BIM communication plan
3.5 To understand procurement strategies for BIM	• Hardware, software and infrastructure services • Procuring specialist BIM services: model production, information management • Injecting BIM requirements into supply chain packages

Index

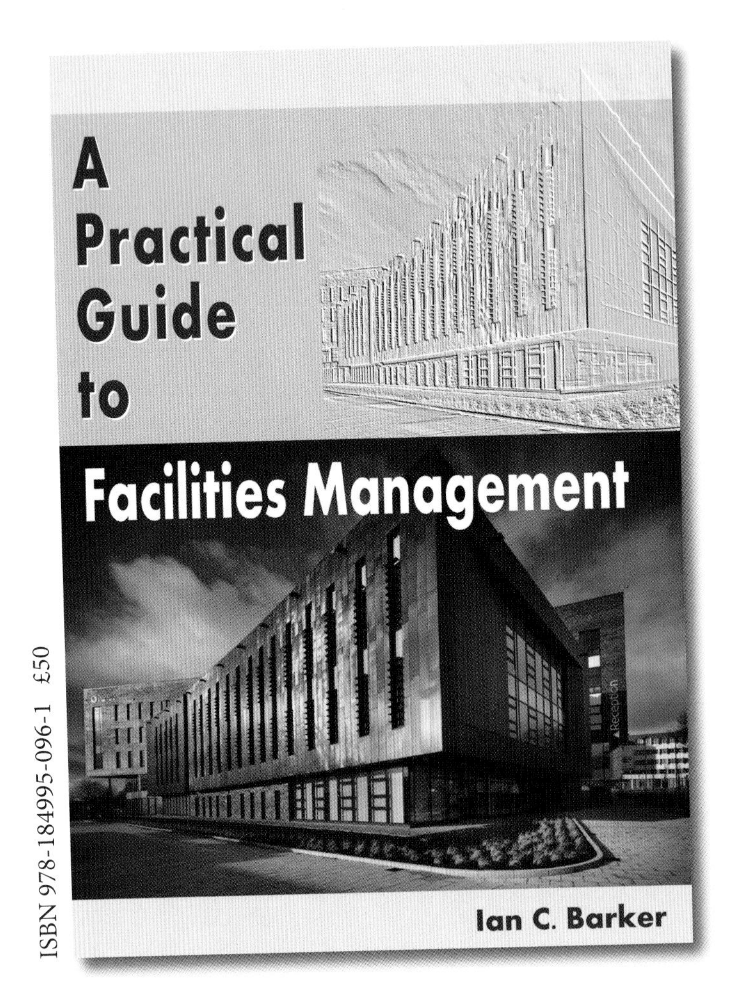

...The detail is bullet point and easy on the reader... I did find myself easily captured by the book and did take out some good points... ...a great read... ***Building Engineer***

...the author takes a practical rather than academic approach to the multiple challenges and constantly moving requirements facing facilities managers. ...is an easy read and provides a quick overview of some FM related topics... ...it has relevance for students and junior facilities managers. ***FMANZ Newsletter***

available from
www.whittlespublishing.com